April 2005

# TECHNOLOGY ASSESSMENT

## Protecting Structures and Improving Communications during Wildland Fires

**GAO**

Accountability ★ Integrity ★ Reliability

GAO-05-380

### GAO
Accountability · Integrity · Reliability

# Highlights

Highlights of GAO-05-380, a report to congressional requesters

# TECHNOLOGY ASSESSMENT

## Protecting Structures and Improving Communications during Wildland Fires

## Why GAO Did This Study

Since 1984, wildland fires have burned an average of more than 850 homes each year in the United States and, because more people are moving into fire-prone areas bordering wildlands, the number of homes at risk is likely to grow. The primary responsibility for ensuring that preventive steps are taken to protect homes lies with homeowners and state and local governments, not the federal government. Although losses from wildland fires made up only 2 percent of all insured catastrophic losses from 1983 through 2002, fires can result in billions of dollars in damages.

Once a wildland fire starts, various parties can be mobilized to fight it, including federal, state, local, and tribal firefighting agencies and, in some cases, the military. The ability to communicate among all parties—known as interoperability—is essential but, as GAO has reported previously, is hampered because different public safety agencies operate on different radio frequencies or use incompatible communications equipment.

GAO was asked to assess, among other issues, (1) measures that can help protect structures from wildland fires, (2) factors affecting use of protective measures, and (3) the role technology plays in improving firefighting agencies' ability to communicate during wildland fires.

www.gao.gov/cgi-bin/getrpt?GAO-05-380.

To view the full product, including the scope and methodology, click on the link above. For more information, contact Robin Nazzaro at (202) 512-3841 or Keith Rhodes at (202) 512-6412.

## What GAO Found

The two most effective measures for protecting structures from wildland fires are: (1) creating and maintaining a buffer, called defensible space, from 30 to 100 feet wide around a structure, where vegetation and other flammable objects are reduced or eliminated; and (2) using fire-resistant roofs and vents. In addition to roofs and vents, other technologies—such as fire-resistant windows and building materials, chemical agents, sprinklers, and geographic information systems mapping—can help in protecting structures and communities, but they play a secondary role.

Although protective measures are available, many property owners have not adopted them because of the time or expense involved, competing concerns such as aesthetics or privacy, misperceptions about wildland fire risks, and lack of awareness of their shared responsibility for fire protection. Federal, state, and local governments, as well as other organizations, are attempting to increase property owners' use of protective measures through education, direct monetary assistance, and laws requiring such measures. In addition, some insurance companies have begun to direct property owners in high-risk areas to take protective steps.

Existing technologies, such as audio switches, can help link incompatible communication systems, and new technologies, such as software-defined radios, are being developed following common standards or with enhanced capabilities to overcome incompatibility barriers. Technology alone, however, cannot solve communications problems for those responding to wildland fires. Rather, planning and coordination among federal, state, and local public safety agencies is needed to resolve issues such as which technologies to adopt, cost sharing, operating procedures, training, and maintenance. The Department of Homeland Security is leading federal efforts to improve communications interoperability across all levels of government. In addition to federal efforts, several states and local jurisdictions are pursuing initiatives to improve communications interoperability.

Source: Larry Korhnak.

Before and after photos illustrating measures to protect a home from wildland fire.

# Contents

## Abbreviations

| | |
|---|---|
| DHS | Department of Homeland Security |
| DOD | Department of Defense |
| GIS | geographic information systems |
| MAFFS | Modular Airborne Fire-Fighting System |
| NAS | National Academy of Sciences |
| NICC | National Interagency Coordination Center |
| NIFC | National Interagency Fire Center |
| SAFECOM | Wireless Public Safety Interoperable Communications Program |
| UHF | ultrahigh frequency |
| VHF | very high frequency |

**United States Government Accountability Office**
**Washington, D.C. 20548**

April 26, 2005

The Honorable Pete V. Domenici
Chairman
The Honorable Jeff Bingaman
Ranking Minority Member
Committee on Energy and Natural Resources
United States Senate

The Honorable Duncan L. Hunter
Chairman
Committee on Armed Services
House of Representatives

The Honorable Vernon J. Ehlers
Chairman
The Honorable David Wu
Ranking Minority Member
Subcommittee on Environment, Technology, and Standards
Committee on Science
House of Representatives

The Honorable Mark Udall
House of Representatives

Consistent with guidance in the Senate's and House's Fiscal Year 2004
Legislative Branch Appropriations Reports (Senate Report 108-88 and
House Report 108-186, respectively), you asked us to conduct an
assessment of technologies used for protecting structures from and
improving communications during wildland fires. This report discusses
measures, including technologies, which can help protect structures from
wildland fires; factors that affect the use of these protective measures; and
the role that technology plays in improving firefighting agencies' ability to
communicate during wildland fires. In addition, appendix I discusses the
process for using military resources in responding to wildland fires.

We are sending copies of this report to the Secretaries of Agriculture,
Commerce, Defense, Homeland Security, and the Interior, as well as to
interested congressional committees. We also will make copies available to
others upon request. In addition, this report will be available at no charge
on the GAO Web site at http://www.gao.gov.

If you have questions concerning this report, please contact Robin Nazzaro at (202) 512-3841 or nazzaror@gao.gov or Keith Rhodes at (202) 512-6412 or rhodesk@gao.gov.

Robin M. Nazzaro
Director, Natural Resources and Environment

Keith A. Rhodes
Chief Technologist
Director, Center for Technology and Engineering

# Executive Summary

Since 1984, wildland fires have burned an average of 850 homes each year in the United States, according to the National Fire Protection Association, but because more people are moving to areas in or near fire-prone wildlands, the number of homes at risk is likely to grow. Such areas, where structures and other human development meet or intermingle with wildlands, are commonly referred to as the wildland-urban interface. In California alone, 3.2 million homes located in the wildland-urban interface are at significant risk from wildland fire.[1] When a large high-intensity wildland fire burns into the wildland-urban interface, it can threaten hundreds of homes at the same time and overwhelm available firefighting resources. Homeowners and state and local governments have the primary responsibility for ensuring that preventive steps are taken to help protect homes from wildland fires, but this does not always happen.[2] While the federal government does not have a primary responsibility, it has played a role through efforts to educate and assist communities in taking preventive steps. Although wildland fires made up only 2 percent of all insured catastrophic losses from 1983 through 2002, the damage from these fires can be costly. For example, wildland fires in Southern California in 2003 caused estimated insured losses of more than $2 billion.

Once a wildland fire starts, many different groups can be mobilized to fight it, including the Forest Service (within the Department of Agriculture); land management agencies in the Department of the Interior; state forestry agencies; local fire departments; private contract firefighting crews; and, in some cases, the military. With many agencies working together, effective communication is essential to fight the fires successfully and to ensure firefighter safety. The ability to communicate among all parties is known as communications interoperability. However, as GAO previously reported, personnel from firefighting and other public safety agencies responding to a fire have had problems communicating with one another

---

[1] California Department of Forestry and Fire Protection, *The Changing California: Forest and Range Assessment 2003* (Sacramento, Calif.: 2003).

[2] In addition to homes, other structures including multiple family dwellings and commercial properties are also threatened by wildland fires. Throughout this report, the terms homes and homeowners refer also to these other structures and property owners.

because agencies operate on different radio frequencies or use different and, sometimes incompatible, communications equipment.[3]

In this context, GAO's review focused on the following issues: (1) measures that can help protect structures from wildland fires, (2) factors affecting the use of these protective measures, and (3) the role that technology plays in improving firefighting agencies' ability to communicate during wildland fires. This report does not discuss fire suppression technologies because it was outside the scope of the study. In addition, GAO was asked to describe the process for using military resources in responding to wildland fires, and this information appears in appendix I.

To obtain information on technologies and other ways for protecting structures from wildland fires, on the factors affecting the use of these measures, and on technologies and other ways for improving communications among agencies fighting wildland fires, GAO worked with the National Academy of Sciences (NAS) to convene a panel of experts for a 2-day symposium in August 2004. GAO also visited six states (California, Florida, Idaho, Montana, New Mexico, and Washington) and met with state and local firefighting or other officials to discuss efforts to protect structures and improve communications when responding to fires, as well as the use of military assistance for firefighting. We selected these states to evaluate a variety of approaches used in different regions of the country with disparate population densities and with varied terrain and vegetation, which can affect the severity of wildland fires. In addition, GAO reviewed studies and pertinent documents and interviewed officials with federal, state, and local agencies and organizations involved in fire research, prevention, and suppression. These organizations included the Forest Service, the Department of the Interior, the Department of Defense, the Department of Homeland Security, and the National Fire Protection Association. Chapter 1 describes GAO's complete scope and methodology.

We conducted our review in accordance with generally accepted government auditing standards from May 2004 to April 2005.

---

[3]See GAO, *Homeland Security: Federal Leadership and Intergovernmental Cooperation Required to Achieve First Responder Interoperable Communications*, GAO-04-740 (Washington, D.C.: July 20, 2004).

# Background

On average, 100,000 wildland fires are reported each year. Firefighting agencies succeed in suppressing more than 95 percent of these fires during initial suppression efforts. But fires that escape initial suppression can grow into large, high-intensity fires burning hundreds of thousands of acres and destroying homes. Under adverse weather and fuel conditions, wildland fires can be difficult to suppress or may be too dangerous to suppress until weather conditions change. Even when firefighters attempt fire suppression, a high-intensity fire in the wildland-urban interface may threaten hundreds of homes simultaneously and overwhelm the firefighting resources available to protect them, as happened during fires in Southern California in 2003. From 2000 through 2003, these suppression efforts cost federal agencies an average of more than $1.3 billion annually.[4]

Recognizing that during severe wildland fires, suppression efforts alone cannot protect all homes threatened by wildland fire, firefighting agencies and others are increasing their emphasis on preventive approaches that help reduce the chance that wildland fires will ignite homes and other structures. Because the vast majority of structures damaged or destroyed by wildland fires are located on private property, the primary responsibility for taking adequate steps to minimize or prevent damage from a wildland fire rests with the property owner and with state and local governments that can establish building requirements and land-use restrictions.

To be able to take effective steps to minimize or prevent damage requires an understanding of the different types of wildland fire and how they can ignite homes.

- *Surface fires* burn vegetation or other fuels near the surface of the ground, such as shrubs, fallen leaves, small branches, and roots.

- *Crown fires* burn the tops, or crowns, of trees. Crown fires normally begin as surface fires and move up the trees by burning "ladder fuel," such as nearby shrubs or low tree branches.

- *Spot fires* are new fires that are started away from the main fire by embers known as "firebrands." Depending on wind conditions, firebrands can be carried a mile or more away from an existing fire.

---

[4]These figures have been adjusted for inflation with fiscal year 2004 as the base year.

Each type of wildland fire threatens structures in different ways. Surface fires can ignite a home or other building by burning nearby vegetation and eventually igniting flammable portions of it, including exterior walls or siding; attached structures, such as a fence or deck; or other flammable materials close by, such as firewood or patio furniture. Crown fires place homes at risk because they create intense heat, which can ignite portions of structures even without direct contact from flames. Firebrands can ignite a structure by landing on the roof or by entering a vent or other opening. Figure 1 illustrates how each type of fire can take advantage of a structure's vulnerabilities and those of its immediate surroundings.

**Figure 1: Ways Wildland Fire Can Threaten a Structure**

Wildland fires can ignite homes and structures in different ways.

**Surface fires (A)** can ignite a home by burning nearby vegetation.

**Crown fires (B)** create intense heat, which can ignite portions of structures, without direct contact from flames.

**Spot fires (C)** started by firebrands can ignite a home by landing on the roof or entering a roof vent or other opening.

Source: GAO.

In responding to a wildland fire, federal, state, local, and tribal firefighting agencies, as well as contractors or the military, may provide personnel and equipment. To help ensure both effective and safe firefighting efforts, firefighters from different agencies need to be able to communicate with one another; that is, they need communications interoperability. During early firefighting efforts, if a number of different firefighting agencies respond to the fire, communications interoperability can become more difficult because these agencies may operate in different bands of the radio frequency spectrum and use equipment that is incompatible.[5]

## Results in Brief

The two most effective measures for protecting structures from wildland fires are: (1) creating and maintaining a buffer around a structure—often called defensible space—by eliminating or reducing trees, shrubs, and other flammable objects within an area from 30 to 100 feet around the structure and (2) using fire-resistant roofs and vents. Analysis of past fires and experimental research have shown that reducing vegetation and other flammable objects within a radius of 30 to 100 feet around a structure, depending on the terrain and vegetation, removes fuels that could bring fire in contact with the structure's walls and can reduce heat generated by a crown fire, which could otherwise damage the structure. Using currently available fire-resistant roof-covering materials, such as asphalt composition shingles rather than untreated wood shingles and screening vents and other openings reduces the likelihood of firebrands igniting a structure. Other technologies can also help in protecting structures and communities, but they play a secondary role. Fire-resistant windows, building materials, and sprinklers that help reduce vulnerability to damage from wildland fire, and technologies such as chemical agents (gels and foams) that coat structures with a temporary protective layer, can also assist in protecting individual homes. In addition, mapping technologies play a supporting role in reducing risk to entire communities. For example, some states and communities use geographic information systems (GIS) mapping to identify and examine the location of structures, fuel distribution, and topography to protect high-risk areas and assist with fire prevention efforts. Two emerging technologies, fire behavior modeling and automated fire detection systems, may also prove useful in the future to protect communities from wildland fires.

---

[5]According to officials, problems with communication occur primarily during the early stages, called the initial and extended attack phases, of an incident. Interoperable radios from national, state, and regional caches can be deployed for large fire operations.

Although protective measures are available, many homeowners do not use them for a number of reasons—including the time or expense involved, competing concerns, misperceptions about how wildland fires ignite structures, and not being aware of their shared responsibility for fire protection—but efforts to increase their use are under way. Fire officials and researchers have reported that some homeowners are discouraged by the time and expense of undertaking protective measures or are reluctant to do so because of concerns over aesthetics or privacy. Officials also said that some homeowners do not recognize the effectiveness of protective measures, such as creating defensible space. Numerous organizations— including federal, state, and local government agencies and nongovernmental organizations—are working to increase the use of measures to protect structures. Some of these efforts seek to increase the voluntary use of protective measures, such as the Firewise Communities program, sponsored by federal agencies and other organizations, that educates homeowners about steps they can take. Other programs directly assist homeowners in creating defensible space. Some jurisdictions have begun to require the use of protective measures. For instance, some state and local governments have adopted laws requiring that homeowners create defensible space around their homes or that homebuilders construct homes and design communities to reduce the risk from wildland fire. Fire officials told GAO, however, that such laws are not always enforced, limiting their effectiveness. Finally, while the insurance industry has not placed a high priority on this issue in the past, some insurance companies direct homeowners in high-risk areas to create defensible space.

While a variety of communications technologies exist to aid interoperability in the short-term—by linking incompatible communication systems used by firefighting and other public safety agencies, commonly called patchwork interoperability—and other technologies are under development to upgrade communications systems to provide increased interoperability in the long term, technology alone cannot solve the interoperability problem. Effective adoption of any of these technologies, whether patchwork solutions, such as audio switches or crossband repeaters, to allow agencies to improve interoperability using existing radio systems, or longer-term system upgrades with radios meeting common standards or utilizing emerging technology, such as software-defined radios, requires planning and coordination among federal, state, and local agencies that work together to respond to wildland fires and other emergencies. Without effective planning and coordination, new investments in communications equipment or infrastructure may not improve communications interoperability among agencies. The

Department of Homeland Security (DHS) is leading federal efforts to improve communications interoperability across all levels of government but, as GAO reported in April 2004, has made limited progress toward achieving interoperability among first responders. Further, GAO reported in July 2004 that DHS does not have the nationwide data necessary to assess interoperability. In that report, GAO recommended that DHS take a variety of actions, including developing a nationwide database and common terminology for public safety interoperability communications channels, to enhance communications interoperability nationwide. DHS agreed with those recommendations. According to a DHS official, as of March 2005, work is under way to develop baseline data. In addition to federal efforts, several states and local jurisdictions are pursuing initiatives to improve communications interoperability.

## Principal Findings

## Defensible Space and Fire-resistant Roofs and Vents Are Key to Protecting Structures; Other Technologies Can Also Help

Managing vegetation and reducing or eliminating flammable objects within 30 to 100 feet[6] of a structure is a key protective measure. Creating such defensible space offers protection by breaking up continuous fuels that could otherwise allow a surface fire to contact and ignite a structure (see fig. 2). Defensible space also offers protection against crown fires. Reducing the density of large trees around structures decreases the intensity of heat from a fire, thus preventing or reducing damage to structures. Analysis of homes burned during wildland fires has also shown defensible space to be a key determinant of whether a home survives. For instance, the 1981 Atlas Peak Fire in California damaged or destroyed 91 out of 111 structures that lacked adequate defensible space but only 5 structures out of 111 that had it. A series of experiments has shown that defensible space can effectively reduce damage to structures from intense crown fires. During these experiments, walls located 33 feet from the crown fires ignited during three of the seven experimental fires and significantly scorched the other four cases. No ignition or observable damage occurred on walls located 66 feet from these crown fires.

---

[6]The distance needed depends on a number of factors, including terrain and vegetation, which can affect fire behavior.

**Figure 2: Home with Defensible Space**

**Defensible space:** Reducing vegetation and other flammable materials within 30 to 100 feet of a structure (the area shaded in yellow) creates defensible space that substantially reduces the likelihood that a wildland fire will damage or destroy the structure. Creating defensible space around a structure does not require that all trees and plants be eliminated, but plants or trees adjacent to structures should be carefully spaced and be pruned to remove the lower branches that hang over the roof.

Source: GAO.

The use of fire-resistant roofs and vents is also important in protecting structures from wildland fires. Many structures are damaged or destroyed by firebrands, which may have traveled a mile or more from the main fire. Fire-resistant roofing materials, such as asphalt composition instead of untreated wood shingles, can reduce the risk that these firebrands will ignite a roof, and vents can be screened with mesh to prevent firebrands from entering and igniting attics. Combining fire-resistant roofs and vents with the creation of defensible space is particularly effective because together these measures reduce the risk from surface fires, crown fires, and firebrands. Studies of two California fires—the 1961 Belair-Brentwood Fire and the 1990 Painted Cave Fire—showed that homes with a nonflammable roof and at least 30 feet of vegetation clearance had more than an 85 percent chance of surviving without active fire protection from firefighters. More recently, California officials attributed one county's success in averting home losses during the 2003 Simi Fire to county laws requiring both fire-resistant roofs and defensible space.

Other technologies play a secondary role in protecting structures from wildland fires. Installing double-paned windows and using fire-resistant materials for siding, for instance, can help reduce risk to structures. Homeowners can obtain additional protection by applying chemical agents, such as gels and foams, to coat the structure with a water-retaining protective layer before a fire arrives. Mapping technologies are also available to improve protection of communities. Florida, for example, has used GIS technology to map and assess the wildland fire risk faced by communities in the wildland-urban interface. Finally, fire officials told GAO that emerging technologies, such as fire behavior modeling and automated detection systems, may prove useful in the future for planning and protecting communities from wildland fires.

## Time, Expense, and Other Competing Concerns Limit the Use of Protective Measures for Structures, but Efforts to Increase Their Use Are Under Way

Many homeowners have not used protective measures—such as creating and maintaining defensible space—because of the time or expense involved in doing so. State and local fire officials estimate that the price of creating defensible space can range from negligible, in cases where homeowners perform the work themselves, to $2,000 or more. Moreover, defensible space needs to be maintained, resulting in additional effort or expense in the future. Competing concerns also influence the use of protective measures. For example, although modifying landscaping to create defensible space has proven to be a key element in protecting structures from wildland fire, officials and researchers have reported that some homeowners are more concerned about the effect landscaping has on the appearance of their property, their privacy, and wildlife habitat. Defensible space, however, can be created in a manner that alleviates many of these concerns. Leaving thicker vegetation away from a structure and pruning plants that remain close to the structure, for instance, can help protect structures from wildland fire and allow them to still be attractive and provide privacy and wildlife habitat.

Misconceptions about fire behavior and the effectiveness of protective measures can also influence the use of steps to protect structures from wildland fires. Fire officials and researchers told GAO that some homeowners do not recognize that a structure and its surroundings constitute fuel that contributes to the spread of wildland fire or understand exactly how a wildland fire ignites structures and, therefore, may not recognize they can take effective steps to reduce their risk. For example, an expert at the symposium convened for GAO by NAS said many homeowners think of wildland fires as intense crown fires and do not believe that any action they take can protect their homes. Officials said that few people realize that reducing tree density close to a structure can return a wildland fire to the ground, where it is much easier to keep away from structures, or that using fire-resistant roofs and screening attic vents can reduce the risk of firebrands igniting homes. Finally, homeowners may not use protective measures because they believe that fire officials are responsible for protecting their homes and do not recognize that they share in this responsibility.

Federal, state, and local agencies, as well as other organizations, are taking a variety of steps intended to increase the creation of defensible space and the use of fire-resistant roofs and vents. Government agencies and other organizations, for instance, are educating people about the effectiveness of simple steps they can take to reduce the risk to structures. Such efforts also demonstrate that defensible space can be attractive, provide privacy, and improve wildlife habitat. In addition to education, some federal, state, and local agencies are directly assisting homeowners in creating defensible space, by providing equipment or financial assistance to reduce fuels near structures. In some cases, government agencies are attempting to further decrease the risk to structures by removing or reducing vegetation in areas immediately adjacent to entire communities. Federal, state, and local agencies, for example, sponsored a project that thinned vegetation to reduce fuels surrounding the town of Roslyn, Washington.

Some state and local governments have adopted laws that require maintaining defensible space around structures or the use of fire-resistant building materials. For example, California requires the creation and maintenance of defensible space around homes and the use of fire-resistant roofing materials in certain at-risk areas. Officials of one county GAO visited attributed the relatively few houses damaged by the 2003 Southern California fires in their county, in part, to its adoption and enforcement of laws requiring defensible space and the use of fire-resistant building materials. Not all states or localities at risk of wildland fire, however, have required such steps. Some state and local officials told GAO that laws had not been adopted because homeowners and developers resisted them. Symposium experts recognized this resistance but emphasized the importance of such state and local laws. Further, to be effective, laws that have been adopted must be enforced, but this does not always happen. Finally, while the insurance industry historically has not placed a high priority on wildland fire issues because of relatively low losses in comparison with other hazards, some insurance companies direct homeowners in high-risk areas to create defensible space.

## Effective Adoption of Technologies to Achieve Communications Interoperability Requires Better Planning and Coordination

Technologies are available or under development to help improve communications interoperability so that personnel from different public safety agencies responding to a fire can communicate effectively. Available technologies include short-term, or patchwork, interoperability solutions to help connect disparate radio systems and allow agencies to use existing communications equipment. One such device is an audio switch that can translate voice or data from one system and make it available and understandable to all other connected communications systems. Other technologies, such as software-defined radios that can transmit and receive a wide range of frequencies, are being developed with enhanced capabilities to overcome interoperability barriers.

Effective adoption of any of these technologies, however, requires planning and coordination among federal, state, and local agencies that work together to respond to emergencies, including wildland fires, to determine the best way to overcome barriers to interoperability. For example, neighboring jurisdictions might choose an interconnection device, such as an audio switch, as a way to improve their communications. To effectively employ the device, they must also jointly decide how to share its cost, ownership, and management; agree on the operating procedures for when and how to deploy it; and train individuals to configure, maintain, and use it. Without such planning and coordination, new investments in communications equipment or infrastructure may not improve the effectiveness of communications among agencies. At the federal level, the

Wireless Public Safety Interoperable Communications Program (SAFECOM) within the Department of Homeland Security is working on a number of initiatives to help state, local, and tribal public safety agencies improve interoperability. An April 2004 GAO report found that limited progress had been made in addressing SAFECOM's overall objective of achieving communications interoperability among entities at all levels of government.[7] Further, a July 2004 GAO report found that nationwide data needed to address the issue of interoperability were not available.[8] In that report, GAO recommended, among other things, that DHS continue to develop a nationwide database and common terminology for public safety interoperability communications channels and assess interoperability in specific locations against defined requirements. DHS agreed with these recommendations. In January 2005, SAFECOM awarded a contract to develop baseline information on the state of interoperability nationwide. In addition to federal efforts, several states and some neighboring local jurisdiction are working to improve interoperability.

## Agency Comments and Our Evaluation

We provided copies of our draft report to the Departments of Agriculture, Commerce, Defense, Homeland Security, and the Interior. The Forest Service, responding for the Department of Agriculture, and the Department of Defense concurred with our report. The Departments of Commerce, Homeland Security, and the Interior generally agreed with our findings but provided technical clarifications on the draft that we incorporated into the report where appropriate. Copies of the written comments from the departments, and our response to them, appear in appendixes VI through X. In addition, we provided copies to the panel of experts that participated in a 2-day symposium convened for GAO by NAS in August 2004. We have incorporated technical and other comments provided by the panelists, as appropriate.

If you have questions about this report, please contact Robin Nazzaro at (202) 512-3841 or nazzaror@gao.gov or Keith Rhodes at (202) 512-6412 or rhodesk@gao.gov. Major contributors to this report are listed in appendix XI.

---

[7]GAO, *Project SAFECOM: Key Cross-Agency Emergency Communications Effort Requires Stronger Collaboration*, GAO-04-494 (Washington, D.C.: April 16, 2004).

[8]GAO-04-740.

# Chapter 1

# Introduction

Fire is a natural process that plays an important role in maintaining the health of many forest and grassland ecosystems, but wildland fire can also endanger the homes and lives of people living in or near wildlands. Areas where structures and other human development meet or intermingle with undeveloped wildland are commonly referred to as the wildland-urban interface. Forest Service and university researchers estimate that more than 42 million homes in the lower 48 states are located in such areas, though the risk from wildland fire in these areas varies widely. When wildland fires threaten homes, personnel and equipment from federal, state, local, or tribal firefighting organizations, as well as contractors or the military, may be mobilized for fire suppression. Effective communication among firefighters and other public safety personnel, primarily using handheld portable radios and mobile radios in vehicles, is needed to ensure safe and successful firefighting efforts.

## Wildland Fires Threaten Homes in Several Ways; Homeowners and State and Local Governments Are Primarily Responsible for Preventive Steps to Protect Them

Although people choosing to live near wildlands may enjoy many benefits from their location, they also run the risk that their homes may be damaged or destroyed by a wildland fire. Wildland fires have destroyed an average of 850 homes per year since 1984, according to a National Fire Protection Association official. However, losses since 2000 have risen to an average of 1,100 homes annually. These losses occurred in many states throughout the nation, including Arizona, California, Florida, and New Mexico, although California has suffered the highest losses overall. Losing homes to wildland fires has long been a problem. Severe fires across the northern United States in 1910 resulted in the destruction of entire towns and, in California, homes have been destroyed in nearly every decade since 1930. The problem is not limited to the United States; wildland fires have damaged or destroyed homes in other countries as well, including Australia, Canada, and France. Most remote wildland fires are ignited by lightning; and humans, intentionally or unintentionally, start the rest.

Fire requires three elements—oxygen, heat, and fuel—to ignite and continue burning. Once a fire has begun, a number of factors—such as terrain, weather, and the type of nearby vegetation or other fuels, including structures—influence how fast and how intensely the fire spreads. For example, fire can burn very rapidly up a steep slope. Adverse weather conditions—especially hot, dry weather with high winds—together with adequate fuels can turn a low-intensity fire into a high-intensity fire that firefighters may be unable to control until the weather changes. Any combustible object in a fire's path, including homes and other structures, can fuel a wildland fire and sustain it. If any one of these three elements is removed, however—such as when firefighters remove vegetation or other fuels from a strip of land near a wildland fire, called a fire break—a fire will normally become less intense and eventually die out.

Wildland fires can threaten homes or other structures in several ways:

- *Surface fires* burn vegetation or other fuels near the surface of the ground, such as shrubs, fallen leaves, small branches, and roots (see fig. 3). These fires can ignite a home by burning nearby vegetation and eventually igniting flammable portions of it, including exterior walls or siding; attached structures, such as a fence or deck; or other flammable materials, such as firewood or patio furniture. (In the electronic version of this report, a video clip illustrating surface fire is available at http://www.gao.gov/media/video/d05380v1.mpg.)

**Figure 3: A Surface Fire**

Source: National Interagency Fire Center.

- *Crown fires* burn the tops, or crowns, of trees. Crown fires normally begin as surface fires and move up the trees by burning "ladder fuel," such as nearby shrubs or low tree branches. Crown fires place homes at risk because they create intense heat, which can ignite portions of structures, if flames are within approximately 100 feet of the structure, even without direct contact. Figure 4 shows a crown fire burning in trees. (In the electronic version of this report, a video clip illustrating crown fire created in an experiment in the Northwest Territories of Canada is available at http://www.gao.gov/media/video/d05380v2.mpg.)

**Figure 4: A Crown Fire**

Source: Forest Service.

- *Spot fires* are started by embers, or firebrands, that can be carried a mile or more away from the main fire, depending on wind conditions. Firebrands can ignite a structure by landing on the roof or by entering a vent or other opening. Firebrands can ignite many homes and surrounding vegetation simultaneously, increasing the complexity of firefighting efforts. (In the electronic version of this report, a video clip illustrating a cloud of firebrands is available at http://www.gao.gov/media/video/d05380v3.mpg.)

Homes can be more flammable than the trees, shrubs, or other vegetation surrounding them (see fig. 5).

**Figure 5: Burning Home Surrounded by Unburned Vegetation**

Source: Forest Service.

Wildland fires can cause extensive and costly damage, but when compared with losses from other natural disasters or even other residential fires, losses from wildland fires are relatively low. From 1983 through 2002, costs and damage from wildland fires in the United States exceeded $1 billion in 2 years and $2 billion in 3 years.[1] During this same 20-year period, however, wildland fires accounted for only about 2 percent of total insured losses from all natural disasters.[2] In contrast, tornadoes accounted for 32 percent

---

[1]National Climatic Data Center, National Oceanic and Atmospheric Administration, *Billion Dollar U.S. Weather Disasters, 1980–2004* (December 2004), http://www.ncdc.noaa.gov (downloaded 1/7/05). According to the report, these cost data include both insured and uninsured losses and were adjusted to 2002 dollars using a gross national product inflation/wealth index.

[2]Insurance Information Institute, *Catastrophes: Insurance Issues,* http://www.iii.org (downloaded 10/15/04).

of total insured losses and hurricanes for 28 percent. In 2003, severe fires in Southern California destroyed more than 3,600 homes, with total damages estimated at more than $2 billion but, in comparison, hurricanes in the Southeast in 2004 damaged an estimated one in five homes in Florida, with estimated total damages of $42 billion. Further, houses damaged or destroyed by wildland fires accounted for less than 1 percent of the estimated 400,000 residential fires that occurred annually from 1994 through 1998.[3]

Losses from wildland fire could increase in the future, as more people move to wildland-urban interface areas. Census Bureau data for 2000 through 2004 indicate that those states with the largest percentage increases in population growth are in the West and South, including Arizona, California, and Florida, where many wildland fires occur. Officials from California, Florida, and New Mexico told us that the wildland-urban interface areas in their states have grown significantly in recent years, and the growth is expected to continue. In California, an estimated 4.9 million of the state's 12 million housing units are located in this area, and 3.2 million of these are at significant risk from wildland fire.[4]

Addressing threats from wildland fires is a shared responsibility. However, homeowners and state and local governments have the primary responsibility for ensuring that preventive steps are taken to help protect homes from wildland fires. While the federal government does not have a primary responsibility, it has played a role through its efforts to educate and assist communities in taking preventive steps. Because the vast majority of structures damaged or destroyed by wildland fires are located on private property, much of the responsibility for taking adequate steps to minimize or prevent damage from wildland fire rests with property owners. State and local governments, as well as the federal government and nongovernmental groups, help to educate homeowners and others about wildland fire and ways to minimize or prevent property damage. State and local officials also can establish and enforce land-use restrictions and laws that require defensible space and fire-resistant building materials. Finally, homebuilders choose the building materials and construction methods

---

[3]Marty Ahrens, *Selections from the U.S. Fire Problem Overview Report: Leading Causes and Other Patterns and Trends: Homes* (Quincy, Mass.: National Fire Protection Association, 2003).

[4]California Department of Forestry and Fire Protection, *The Changing California: Forest and Range 2003 Assessment* (Sacramento, Calif.: 2003).

used, in accordance with local building codes, when building a home, and insurance companies reimburse their clients for losses, including those from wildland fires.

# Multiple Agencies Respond to Wildland Fires and Cannot Always Communicate Effectively with One Another

Once a wildland fire starts, many different agencies assist in the efforts to manage or suppress it. To fight fires, the United States uses an interagency system whereby needed personnel, equipment, aircraft, and supplies are ordered through a three-tiered—local, regional, and national—dispatching system. Federal, state, local, and tribal government agencies; private contractors; and, in some cases, the military, supply firefighting personnel and equipment, which is coordinated through various dispatch centers. The National Interagency Coordination Center (NICC) in Boise, Idaho, is the primary center for coordinating and mobilizing wildland firefighting resources nationwide. NICC is also responsible for coordinating with the Department of Defense (DOD) if military assets are needed. When requests exceed available resources, fires are prioritized, with those threatening lives and property receiving higher priority for resources. Although this interagency response system is an effective way to leverage limited firefighting resources, communications challenges may arise because the various agencies responding to a fire may communicate over different radio frequency bands or with incompatible communications equipment. Problems with communications interoperability occur primarily during the early efforts to suppress the fire, called the initial and extended attack phases, before national and state caches of interoperable radios can be deployed to the incident.

Land mobile radio systems are the primary means of communication among public safety personnel operating in a single area. These systems consist of a regularly interacting set of components including a base station, which controls the transmission and reception of audio signals among radios; mobile radios in vehicles and handheld portable radios carried by emergency personnel; and stations, known as repeaters,[5] which relay radio signals (see fig. 6).

---

[5]Using repeaters increases the distance over which radio users can communicate with one another.

Figure 6: Basic Components of a Land Mobile Radio Communication System

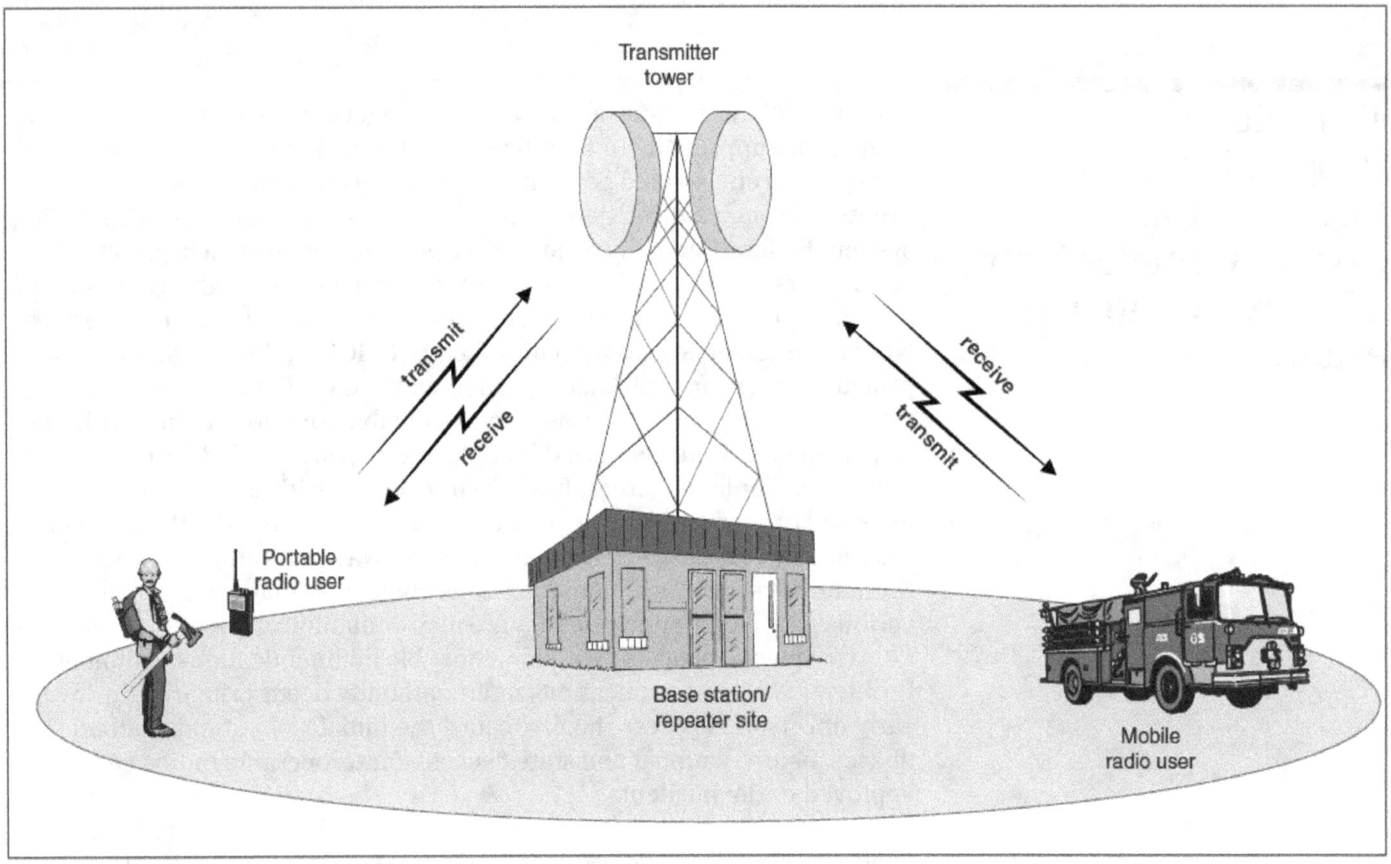

Sources: GAO, Department of Homeland Security, and Nova Development Corp.

Radio signals travel through space in the form of waves. These waves vary in length, and each wavelength is associated with a particular radio frequency.[6] Radio frequencies are grouped into bands. Of the more than 450

___

[6]Radio frequencies are measured in Hertz (Hz); the term *kilohertz* (kHz) refers to thousands of Hertz, *megahertz* (MHz) to millions of Hertz, and *gigahertz* (GHz) to billions of Hertz.

frequency bands in the radio spectrum, 10,[7] scattered across the spectrum, are allocated to public safety agencies (see fig. 7). The radio spectrum is finite, however, and additional frequencies cannot be added or created. As a result, efforts are increasing to make more efficient use of existing spectrum, including moving toward narrowband radios, which use channels 12.5 kHz wide, in contrast to the channels 25 kHz wide used by wideband radios.[8]

Figure 7: Public Safety Agency Radio Frequency Bands and Their Location on the Spectrum

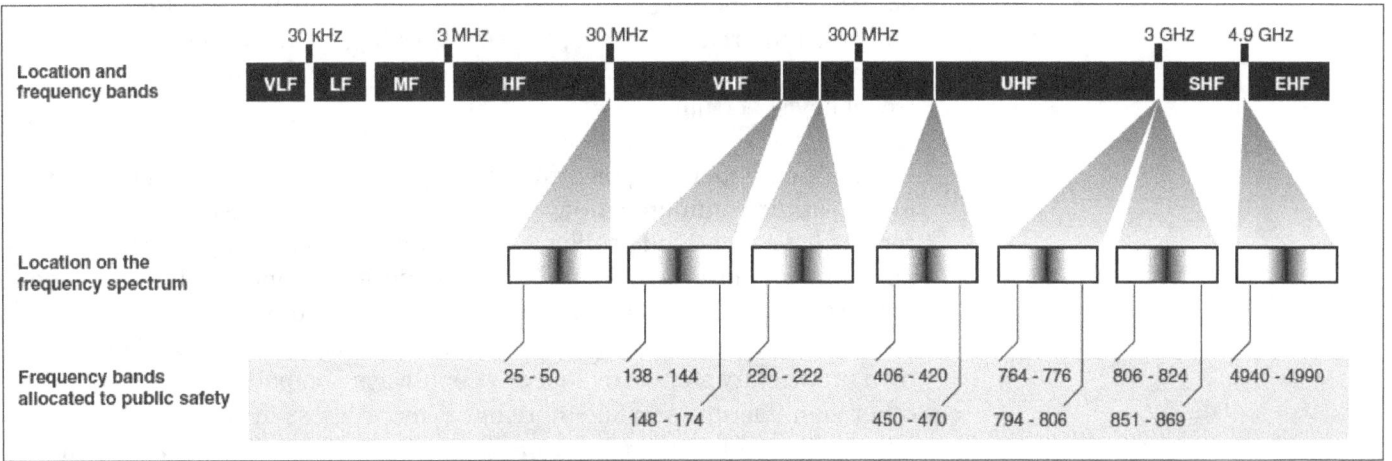

Sources: GAO and Department of Homeland Security.

Note: Federal firefighting agencies primarily operate in the VHF band (162-174 MHz) and the UHF band (406-420 MHz). State and local public safety agencies operate in one or more of the bands depending on their particular needs and circumstances.

---

[7]In addition to the 10 public safety bands, the Federal Communications Commission's allocation of 50 MHz of spectrum in the 4.9 GHz band also provides public safety agencies with the ability to support new broadband applications such as high-speed digital technologies and to implement on-scene wireless networks for activities including transfers of large amounts of data, such as for maps.

[8]Federal and other public safety agencies are adopting narrowband capabilities at different rates. The National Telecommunications and Information Administration, which regulates the federal government's use of the radio spectrum, has mandated that federal agencies generally must adopt narrowband communications capability by 2008. The Federal Communications Commission regulates other public safety agencies in the VHF and UHF bands and does not generally require them to adopt narrowband technology until 2018.

A firefighting or public safety agency typically uses a radio frequency band appropriate for its locale, either rural or urban. Bands at the lower end of the radio spectrum, such as VHF (very high frequency), work well in rural areas where radio signals can travel long distances without obstruction from buildings or other structures. Federal firefighting agencies, such as the Forest Service, and many state firefighting agencies operate radios in the VHF band. In urban areas, firefighting and other public safety agencies may operate radios on higher frequencies, such as those in the UHF (ultrahigh frequency) or 800 MHz bands, because these frequencies can penetrate buildings and provide better communications capabilities for an urban setting. As we previously reported, when federal, state, and local emergency response agencies work together, for example to fight a fire in the wildland-urban interface, they may not be able to communicate with one another because they operate in different bands along the radio frequency spectrum.

In addition to operating on different frequency bands, some agencies use incompatible communications systems that are not interoperable. Various reports have identified problems with agencies using aging or incompatible communications systems as a factor hampering communications between public safety agencies. Incompatible communications systems exist, in part, because some manufacturers make radio equipment based on their own proprietary standards that are not always compatible with those of other manufacturers. While there has been progress in developing national standards to help ensure interoperability, lack of funding can affect an agency's ability to upgrade to newer communications systems based on these standards. The lack of communications interoperability among firefighting and other first-responder agencies can impair their ability to respond to emergencies quickly and safely, and cost lives among responders and those they are trying to assist.

## Objectives, Scope, and Methodology

Our review addressed the following objectives: (1) measures that can help protect structures from wildland fires, (2) factors that affect the use of these protective measures, and (3) the role that technology plays in improving firefighting agencies' ability to communicate during wildland fires. In addition, we were asked to describe the process for using military resources in responding to wildland fires.

To address the first three of these objectives, as detailed below, we contracted with the National Academy of Sciences (NAS) to convene a symposium of experts and we visited six states. In addition, we reviewed studies and other pertinent documents and conducted interviews with a broad range of individuals and organizations to obtain information to address individual objectives.

We conducted our review in accordance with generally accepted government auditing standards from May 2004 to April 2005.

## Symposium Convened for GAO by the National Academy of Sciences

We worked with NAS to convene a panel of experts for a 2-day symposium in August 2004.[9] This symposium addressed the role of technology and other measures to help protect structures from wildland fires and the factors affecting their use. It also addressed technologies for improving communications among agencies fighting wildland fires. Twenty-five experts participated in the symposium. (See app. II for a list of participants.) Federal experts included scientists or specialists in fire behavior, building and materials technologies, and communications technologies. Other experts included county and city firefighting officials, university researchers specializing in behavioral sciences or risk management, and specialists on building codes and other fire protection measures.

## Site Visits to Six States

To obtain additional information on our objectives and to identify different approaches that regions, states, or communities are taking to address the risk to structures from wildland fire, interoperability of communications, or use of military resources, we conducted site visits to six states: California, Florida, Idaho, Montana, New Mexico, and Washington. We selected these states to evaluate a variety of approaches used in different regions of the country with disparate population densities and varied terrain and vegetation, which can affect the severity of wildland fires. At each location, we reviewed documents and interviewed officials to discuss: (1) the steps that can be taken to protect structures from wildland fires, including efforts that encourage the voluntary use of these steps and those requiring their use; (2) the factors affecting the use of these steps; and (3) the status of communications interoperability and efforts being made to address communications difficulties. At each location, we also interviewed

---

[9]We have a standing contract with NAS under which NAS provides assistance in convening groups of experts to provide information and expertise to our engagements. NAS uses its scientific network to identify participants and uses its facilities and processes to arrange the meetings. Recording and using the information in a report is our responsibility.

state and local officials, including fire managers or firefighters, fire marshals, emergency management personnel, elected officials, and other government officials such as land-use planners. In addition, we interviewed homeowners in several of the visited states to obtain their perspective on the effectiveness of measures to protect structures from wildland fires and the efforts to increase use of such measures.

## Additional Efforts to Address Individual Objectives

To gather information on the measures that can help protect structures from wildland fires, we reviewed studies and pertinent documents and interviewed officials with federal agencies involved in fire research, building construction and materials design and research, fire prevention efforts, and fire suppression. Our sources included the Forest Service within the Department of Agriculture and several of its research stations, including the Fire Science Laboratory, the Missoula Technology and Development Center, and the Forest Products Laboratory; the Department of the Interior, including the Bureau of Land Management; the National Institute of Standards and Technology within the Department of Commerce; and the National Interagency Fire Center in Boise, Idaho. We also interviewed representatives from other organizations including the Institute for Business and Home Safety, the National Fire Protection Association, and the National Association of Homebuilders. The scope of our study included technologies that could be incorporated into structures or into communities to help them better withstand wildland fires, but it did not include technologies for the suppression of wildland fires.

To identify factors affecting the use of protective measures and the steps being taken to increase their use, we carried out a number of activities. First, because the primary national effort to reduce fire risk to structures is the Firewise Communities program, we reviewed Firewise Communities program documents and interviewed program officials and a range of program participants. We also attended a 2004 national Firewise Communities conference in Denver, Colorado, which addressed current efforts and remaining challenges, and a 2004 Forest Service conference in Boise, Idaho, which addressed wildland fire issues. Second, we reviewed government and other research studies examining the use of protective measures and the effectiveness of programs designed to increase their use. Third, to expand the geographic coverage of our study and to identify broader concerns, we reviewed documents or interviewed officials from federal firefighting agencies, the Federal Emergency Management Agency within the Department of Homeland Security, the National Association of Counties, and the Western Governors' Association. Finally, to obtain information on the role of the insurance industry in protecting structures from wildland fires, we interviewed officials from the Insurances Services Office,[10] the California FAIR plan program,[11] the Personal Insurance Federation of California,[12] state insurance agencies from several states, and from two insurance companies.

To gather information on the role that technology plays in improving firefighting agencies' ability to communicate during wildland fires, we reviewed reports including previous GAO reports on interoperability and radio spectrum management, National Task Force on Interoperability reports, and Wireless Public Safety Interoperable Communications Program (SAFECOM)[13] reports. We also interviewed officials from federal agencies involved in firefighting, including the Forest Service, the Bureau of Land Management, and the National Interagency Communications Center at the National Interagency Fire Center in Boise, Idaho, and federal agencies involved in communications technologies and related issues, including the Office of the Assistant Secretary of Defense for Homeland Defense and the Naval Research Laboratory, both within the Department of Defense, and the Federal Emergency Management Agency and the Office of Interoperability and Compatibility, both within the Department of Homeland Security. We obtained information on available communications technologies from several manufacturers.

To obtain information on the use of military resources, we reviewed relevant legislation, agreements between DOD and federal or state firefighting agencies, policies, and procedures governing the use of military resources to fight wildland fires. We also reviewed reports evaluating the use of military resources including a 2004 Office of Management and Budget report and reports on the Southern California fires of 2003. We spoke with officials from the Office of the Assistant Secretary of Defense

---

[10]The Insurance Services Office, based in Jersey City, NJ, provides data, analysis, and consulting services to the insurance industry.

[11]According to an official, the state of California established the Fair Access to Insurance Requirements (FAIR) program in 1968 to assist home and business owners who had difficulty obtaining fire insurance.

[12]The Personal Insurance Federation of California is a trade association representing insurance companies that provide 50 percent of personal insurance in California.

[13]SAFECOM is managed by the Department of Homeland Security. Its goal is to achieve interoperability among emergency-response communications at all levels of government.

for Homeland Defense and fire or military officials in California, Florida, Idaho, New Mexico, and Washington to obtain their perspectives on the use of military resources to assist wildland fire suppression efforts in those states.

# Defensible Space and Fire-resistant Roofs and Vents Are Key to Protecting Structures; Other Technologies Can Also Help

Creating and maintaining defensible space and using fire-resistant roofs and vents are critical to protecting structures from wildland fires. Analysis of past fires and research experiments have shown that reducing vegetation and other flammable materials within a radius of 30 to 100 feet[1] around a structure removes fuels that could bring a surface fire in contact with the structure's walls and can reduce heat generated by a crown fire that could otherwise damage the structure. Although defensible space can reduce the risk from surface and crown fires, it cannot prevent firebrands from igniting the roof or entering an opening and igniting a structure. Using fire-resistant roof-covering materials, which inhibit ignition, and screening exterior vents and other openings can help protect against firebrands and provide another important level of protection. Several other technologies can supplement defensible space and fire-resistant roofs and vents. Some of these technologies, like chemical agents, help protect individual structures, while others, like geographic information systems, help protect communities.

## Defensible Space and Fire-resistant Roofs and Vents Are Critical to Protecting Structures

Managing vegetation and reducing or eliminating flammable materials within 30 to 100 feet of a structure creates a defensible space that substantially reduces the likelihood that a wildland fire will damage or destroy the structure. Because wildland-urban interface fires may threaten hundreds of homes simultaneously and overwhelm the firefighting resources available to protect them, the goal of defensible space is to protect a structure from wildland fire without requiring fire suppression.[2] Defensible space offers protection by breaking up continuous fuels (including plants, leaves, needles, or debris) that could otherwise allow a surface fire to contact the structure and ignite it. Defensible space also helps protect against crown fires. Reducing the density of large trees around a structure decreases the heat intensity of any nearby fire, thus helping to prevent structures from igniting.

---

[1]The amount of defensible space needed can be affected by a number of factors, including terrain and vegetation. In certain circumstances, effective defensible space may need to exceed 100 feet. Ventura County, California, for example, recommends that homeowners create 200 feet of defensible space around homes located near the top of a slope, facing east or south, or near heavy chaparral vegetation.

[2]Because of the importance of protecting a structure from wildland fire without requiring fire suppression efforts, some fire officials use other terms including "home ignition zone" or "self-defending space" to refer to this concept.

Chapter 2
Defensible Space and Fire-resistant Roofs
and Vents Are Key to Protecting Structures;
Other Technologies Can Also Help

Defensible space begins at the outer limit of any exterior component of a structure and does not require that all trees and plants be eliminated (see fig. 8). The 30 to 100 feet of defensible space extends beyond exterior components such as decks, fences, or porches and, under certain conditions, homeowners may keep some plants or trees adjacent to their homes. Plants within the 30-to-100-foot radius should be carefully spaced and not highly flammable. Trees should have their lower branches removed, with no branches hanging over the roof. In addition, moving other flammable materials, such as firewood piles and flammable outdoor furniture, away from the structure also contributes to defensible space.

Chapter 2
Defensible Space and Fire-resistant Roofs
and Vents Are Key to Protecting Structures;
Other Technologies Can Also Help

**Figure 8: Home with Defensible Space**

**Defensible space:** Reducing vegetation and other flammable materials within 30 to 100 feet of a structure (the area shaded in yellow) creates defensible space that substantially reduces the likelihood that a wildland fire will damage or destroy the structure. Creating defensible space around a structure does not require that all trees and plants be eliminated, but plants or trees adjacent to structures should be carefully spaced and be pruned to remove the lower branches that hang over the roof.

Source: GAO.

Chapter 2
Defensible Space and Fire-resistant Roofs
and Vents Are Key to Protecting Structures;
Other Technologies Can Also Help

When individual homeowners do not own 30 to 100 feet of property around their homes, as is the case in many subdivisions, homeowners may need to cooperate with neighbors or adjacent property owners to ensure that adequate defensible space is created and maintained across multiple properties. Figure 9 shows a subdivision in California that managed vegetation between homes and around the community and survived a wildland fire in 2004.

Chapter 2
Defensible Space and Fire-resistant Roofs
and Vents Are Key to Protecting Structures;
Other Technologies Can Also Help

**Figure 9: A California Community with Defensible Space That Survived a Wildland Fire in 2004**

Source: Ventura County Fire Department.

In addition to creating and maintaining defensible space, effective wildland fire protection calls for both roofing with fire-resistant materials and screening exterior vents or openings to keep out firebrands, which can travel a mile or more through the air. Although defensible space can reduce the risk from crown and surface fires, it cannot prevent firebrands from entering and igniting a structure's highly flammable interior.

Chapter 2
Defensible Space and Fire-resistant Roofs
and Vents Are Key to Protecting Structures;
Other Technologies Can Also Help

Roofs can be made fire-resistant by using appropriate protective covering materials, either when building new homes or retrofitting or remodeling existing homes. Materials such as asphalt composition, clay, concrete, metal, slate, treated wood products, and even synthetics, such as rubber, can all be used to achieve a "class A" roof.[3] Some of these protective covering materials will not ignite even on direct contact with fire. These fire-resistant covering materials are available at costs similar to more flammable materials, such as cedar shakes.[4] In addition to covering material, a roof's design, construction quality, and condition also influence its susceptibility to ignition. For example, certain complex roof patterns have valleys and crevices that can trap leaves, needles, and other flammable debris, increasing the likelihood of ignition.

Even when defensible space and fire-resistant roofing protect a structure from the outside, it can still ignite from within if firebrands enter through vents or other openings. Most structures have some ventilation in crawl spaces or attics for moisture control (see fig. 10). Often located at the gable[5] ends of the roof or under the eaves,[6] such vents allow air to flow into and through the attic. Other openings may also be left by poor construction, deterioration, or ill-fitting joints between walls and roof. Covering vents and other openings with screens that will not burn or melt, substantially reduces the risk of entry and ignition by firebrands. The Firewise Communities program, a national program which educates

---

[3]Existing standardized tests of fire resistance evaluate entire roof assemblies, rating them class A, B, or C according to tests approved by the American National Standards Institute/Underwriters Laboratories Inc. and the American Society for Testing and Materials. In these tests, burning firebrands of different sizes are placed on top of the roof assembly. Large brands are used to test for a class A rating, smaller brands for class B or C ratings. If the brand does not burn though a roof assembly in 90 minutes, the assembly passes the test for a given rating. Class A roof coverings are considered effective against severe fire exposures. Time-dependent ratings, however, may not be meaningful in a wildland fire scenario because firefighters may not be able to respond for many hours, if at all.

[4]Composition shingles and some metal coverings, for instance, can cost less than wood shingles or wood shakes.

[5]A gable is typically a triangular section of wall at the end of a pitched roof, occupying the space between the two slopes of the roof.

[6]Eaves are the edges or lower borders of a roof overhanging the exterior walls.

Chapter 2
Defensible Space and Fire-resistant Roofs
and Vents Are Key to Protecting Structures;
Other Technologies Can Also Help

homeowners about wildland fire and steps to protect homes against them, recommends screen openings be one-eighth inch or less.[7]

**Figure 10: Roof and Vents**

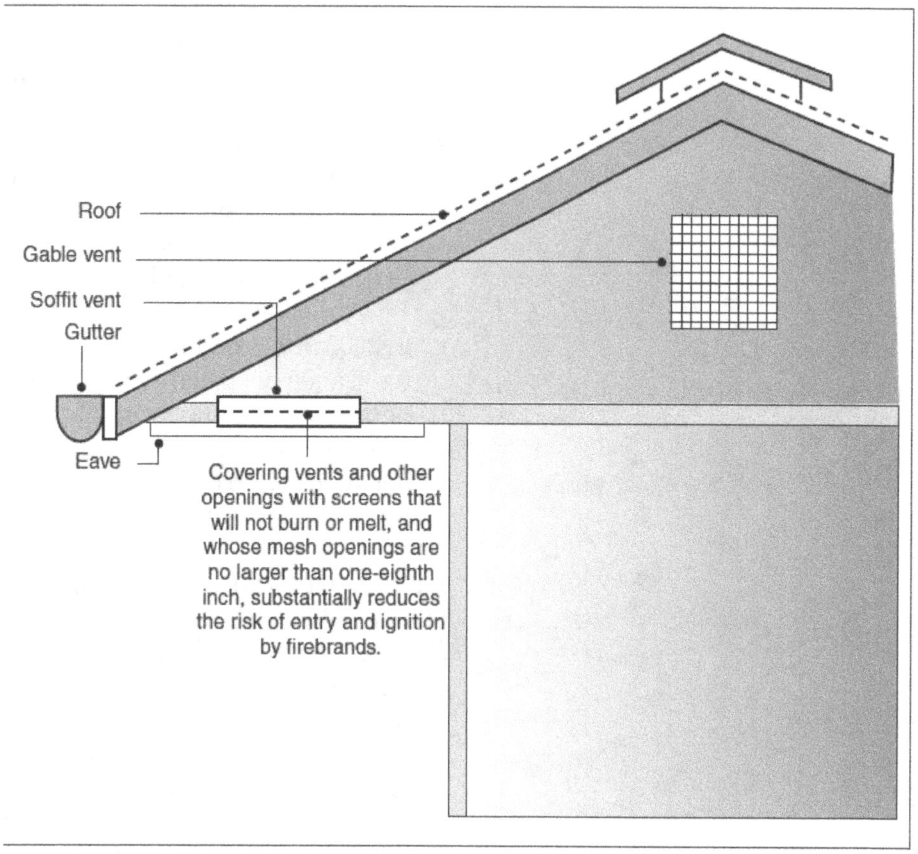

Roof
Gable vent
Soffit vent
Gutter
Eave

Covering vents and other openings with screens that will not burn or melt, and whose mesh openings are no larger than one-eighth inch, substantially reduces the risk of entry and ignition by firebrands.

Sources: Florida Department of Community Affairs, Florida Department of Agriculture and Consumer Services, and GAO.

Analysis of fires over the last half century has demonstrated the importance of defensible space and fire-resistant roofs and vents as protective measures for structures.

---

[7]Some building code guidance recommends screen openings be one-quarter inch or less. However, the Governor's Blue Ribbon Fire Commission established shortly after the 2003 Southern California fires to review the firefighting efforts and recommend improvements, found that one-quarter inch mesh screens were insufficient to prevent entry of firebrands.

Chapter 2
Defensible Space and Fire-resistant Roofs
and Vents Are Key to Protecting Structures;
Other Technologies Can Also Help

- In the 1961 Belair-Brentwood Fire and the 1990 Painted Cave Fire, both in California, 85 to 95 percent of homes with a nonflammable roof, and at least 30 feet of defensible space, survived without fire department intervention.

- In the 1981 Atlas Peak Fire in California, out of the 323 structures threatened, only 5 of the 111 structures with defensible space were damaged or destroyed. In contrast, 91 of the 111 structures without defensible space were either damaged or destroyed.[8]

- In the 1985 Palm Coast Fire in Florida, 130 homes were damaged or destroyed. Two of the most predictive factors for whether homes in this fire burned or survived were fire-resistant vents and defensible space. Those homes with flammable, unprotected vents were identified as particularly vulnerable.

- In 2003, the Simi Fire in Ventura County, California, threatened thousands of structures. According to the Ventura County fire marshal, of the few structures actually destroyed during these fires, most did not observe the county's ordinance requiring 100 feet of defensible space between the structure and flammable vegetation, or they lacked county-recommended fire-resistant roofs and properly screened vents.

Experimental research on wildland fire has corroborated the effectiveness of defensible space and fire-resistant roofs. A researcher at the Forest Service's Fire Science Laboratory in Missoula, Montana, predicted that a crown fire would have to come within 100 feet of a structure for it to ignite; he based this prediction on a theoretical model incorporating conservative estimates of the heat an intense crown fire would produce and the ignitability of wood.[9] The researcher tested the model's results in a series of experiments while working with a group of international fire researchers in Canada's Northwest Territories (see fig. 11). During these experiments, five-and-a-half acre plots of trees were ignited under conditions that produced a crown fire. Wood walls were exposed at varying distances to the fire's heat. Walls located 33 feet from the crown fire ignited during three of seven experimental fires and significantly scorched in the other four

---

[8]The remaining 101 properties did not have adequate defensible space, and about half of the structures were damaged or destroyed.

[9]Jack Cohen, "Preventing Disaster: Home Ignitability in the Wildland-Urban Interface." *Journal of Forestry* 98 (2000): 15-21.

Chapter 2
Defensible Space and Fire-resistant Roofs
and Vents Are Key to Protecting Structures;
Other Technologies Can Also Help

fires. Walls located 66 feet from the crown fire did not ignite or sustain visible damage. These experiments also demonstrated that fire-resistant roofs can effectively protect structures' highly flammable interiors from igniting. Using a model structure with a roof covering made from composition shingles, fire researchers also set fire to the pine needles completely covering the roof. The composition roof did not ignite, and the structure remained undamaged. (In the electronic version of this report, a video clip illustrating this experiment is available at http://www.gao.gov/media/video/d05380v4.mpg.)

**Figure 11: Fire Experiments in Canada's Northwest Territories**

Source: Forest Service.

Finally, experts from the symposium convened for us by the National Academy of Sciences (NAS) emphasized that defensible space and fire-resistant roofs and vents are the most critical protective measures. Symposium experts stated that defensible space is critical for protecting structures from wildland fire. These experts told us that if defensible space and fire-resistant roofs and vents were correctly and consistently used by homeowners, the risk posed by wildland fire would be significantly

Chapter 2
Defensible Space and Fire-resistant Roofs
and Vents Are Key to Protecting Structures;
Other Technologies Can Also Help

reduced. Moreover, in visits to California, Florida, Idaho, Montana, New Mexico, and Washington, we met with fire officials who confirmed the symposium experts' view—that 30 to 100 feet of defensible space and fire-resistant roofs and vents are vital to protecting structures from wildland fires.

## Other Technologies Play a Secondary Role

Symposium experts and fire officials we spoke with identified other technologies that can help protect individual structures from wildland fires. A few of these technologies, like fire-resistant building materials (other than roofing), are permanent, requiring little intervention by homeowners or firefighters, while other technologies, like chemical agents, are temporary and require active human intervention. Still other technologies, like geographic information systems (GIS) mapping, can be used to help protect entire communities. See appendix III for more information on these technologies.

- *Fire-resistant windows.* Fire-resistant windows help protect a structure from wildland fire by reducing the risk a window will break and allow fire to enter a structure. Windows constructed of double-paned glass, glass block, or tempered glass can help resist breakage.

- *Fire-resistant building materials.* Fire-resistant building materials for walls, siding, decks, and doors play an important role in protecting structures by helping to prevent ignition. During a wildland fire, flames or firebrands may come in contact with a structure or intense heat may either ignite the exterior of a structure or melt it, thus exposing the structure's interior to the fire. Exterior walls, siding, decks, and doors made of fire-resistant building materials, such as fiber-cement, brick, stone, metal, and stucco, help structures resist such damage and destruction.

- *Chemical agents.* Firefighting chemical agents, such as foams and gels, are temporary protective measures that can be applied as an exterior coating shortly before a wildland fire reaches a structure. Foams, typically detergent based, are combined with water or forced air. Gels are polymers (plastics) that can hold many times their weight in water. Both are designed to be sprayed onto a structure, coating it with a protective outer shield against ignition (see fig. 12). For example, California Division of Forestry and Fire Protection officials estimated that in 2003, using gels helped save between 75 and 100 homes from the Paradise Fire and more than 300 homes from the Cedar Fire in San

Chapter 2
Defensible Space and Fire-resistant Roofs
and Vents Are Key to Protecting Structures;
Other Technologies Can Also Help

Diego County. The disadvantages of using foams and gels are that they often need to be applied to a structure by a homeowner or firefighter. Chemical agents may also need to be periodically reapplied or sprayed with water to remain effective, and they can be difficult to clean up.

**Figure 12: Firefighter Applying a Chemical Agent to a Home**

Source: Thermo Technologies.

- *Sprinkler systems.* Sprinkler systems, which can be installed inside or outside a structure, lower the risk of ignition or damage. For example, the California Governor's Blue Ribbon Commission recommended adding internal attic sprinklers to revised building codes as a response to lessons learned from the 2003 wildland fires. Sprinklers, however, require reliable sources of water and, in some cases, electricity to be effective. Several firefighting officials told us that during wildland fires,

Chapter 2
Defensible Space and Fire-resistant Roofs
and Vents Are Key to Protecting Structures;
Other Technologies Can Also Help

power and water services may not be adequate for sprinklers to function properly. For example, an investigation after California's 1991 Oakland Hills Fire noted that external sprinkler systems might have saved some homes if water flow and pressure had been adequate.

In addition to technologies aimed at protecting individual structures, symposium experts and fire officials we met with told us that one important technology exists, geographic information systems (GIS) mapping, that can reduce the risk of wildland fire damage to an entire community. GIS is a computer-based information system that can be used to efficiently store, analyze, and display multiple forms of information on a single map.[10] GIS technologies allow fire officials and local and regional land managers to combine vegetation, fuel, and topography data into separate layers of a single GIS map to identify areas in need of vegetation management or to set priorities for fuel breaks. State and county officials we met with emphasized the value of GIS in community education and community-planning efforts to protect structures and communities from wildland fire damage within their jurisdictions. For example, the state of Florida has developed the Florida Risk Assessment System. This interactive GIS provides Florida Division of Forestry officials a detailed visual representation of data on fuels, topography, and weather. Displaying these data on one map helps officials determine which communities are at high risk and identify which areas near these communities need treatments to reduce fuels (see fig. 13).[11]

---

[10]GIS also has applications related to wildland fire suppression activities, including preplanning for evacuations during wildland fires. For additional information on how GIS can assist wildland fire management, see: GAO, *Geospatial Information: Technologies Hold Promise for Wildland Fire Management, but Challenges Remain,* GAO-03-1047 (Washington, D.C.: Sept. 23, 2003).

[11]The Forest Service and Department of the Interior are currently developing a national data and modeling GIS system, called LANDFIRE. More information on LANDFIRE can be found in GAO, *Wildland Fire Management: Important Progress Has Been Made, but Challenges Remain to Completing a Cohesive Strategy,* GAO-05-147 (Washington, D.C.: Jan. 14, 2005).

Chapter 2
Defensible Space and Fire-resistant Roofs
and Vents Are Key to Protecting Structures;
Other Technologies Can Also Help

**Figure 13: GIS Map Showing Levels of Concern in Myakka River District, Florida**

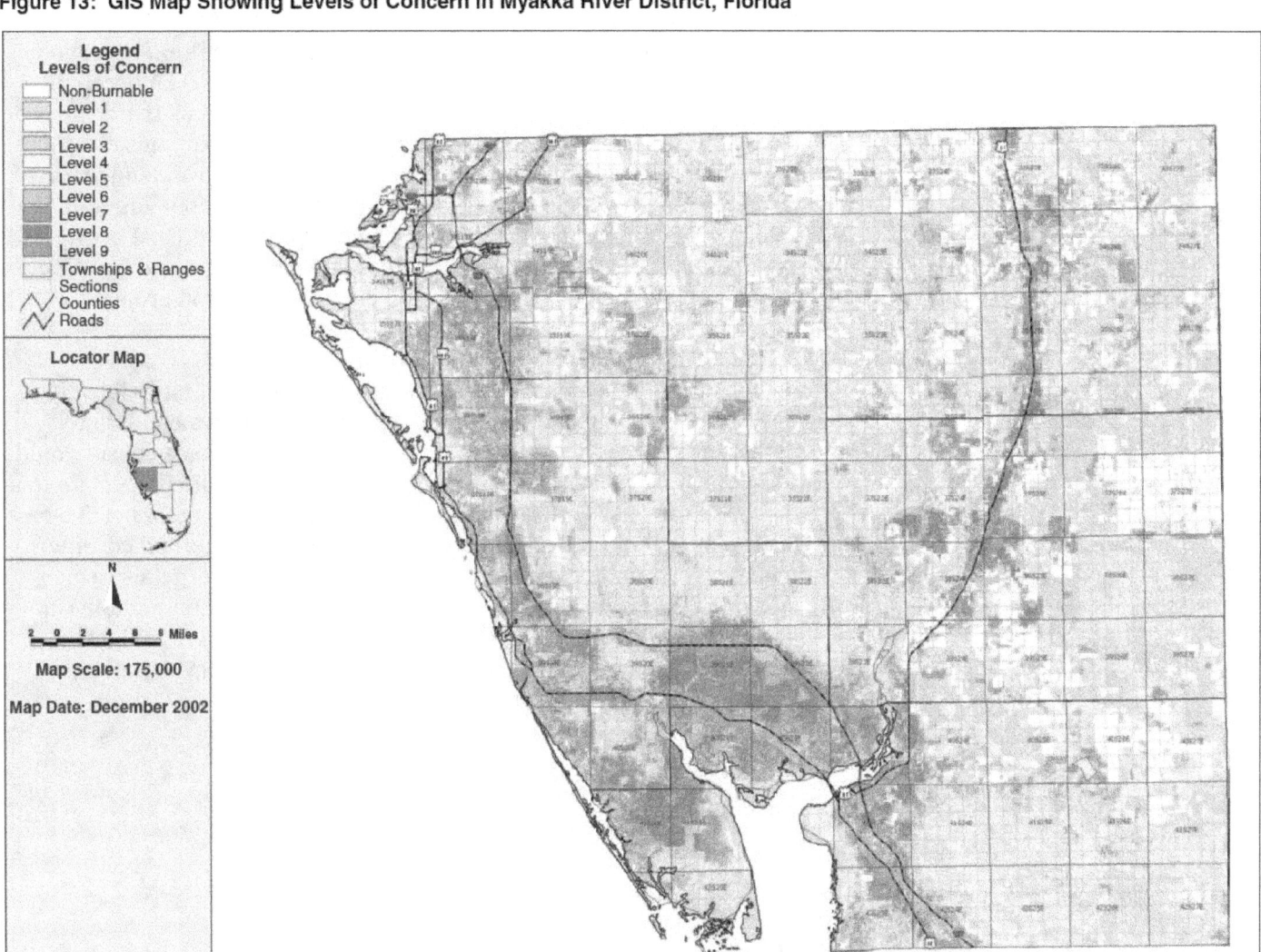

Source: Florida Division of Forestry.

Some emerging technologies could assist in protecting communities, although they need more research, testing, and time to fully develop. Emerging technologies are as follows:

- *Fire behavior modeling.* Forest Service and other researchers have developed computer models to predict wildland fire behavior, but these models do not accurately predict fire behavior in the wildland-urban

Chapter 2
Defensible Space and Fire-resistant Roofs
and Vents Are Key to Protecting Structures;
Other Technologies Can Also Help

interface. Existing models have helped officials identify areas likely to experience intense wildland fires, identify suitable locations for fuel breaks, predict the effect of a fuel break on fire behavior, and aid suppression by predicting overall behavior of a given fire. These models do not, however, consider the effect that structures and landscaping have on wildland fire behavior. Some researchers told us that developing models that consider how fire spreads from house to house might help improve the design of communities in the wildland-urban interface. Such models might also help homeowners compare how different landscaping options could alter fire behavior. The Forest Service, National Institute of Standards and Technology, and Los Alamos National Laboratory have proposed a 5-year, $10 million project to develop such models.

- *Automated detection systems.* Sensors using infrared, ultraviolet, or temperature-sensitive devices[12] can be placed around a community[13] to detect the presence of wildland fire. On detecting a fire, a sensor could set off an audible alarm or could be connected via radio or satellite to a device that would notify homeowners or emergency personnel. Several such sensors could be networked together to provide broad coverage of the area surrounding a community. According to fire officials, sensor systems may prove particularly helpful in protecting communities in areas of rugged terrain or poor access where wildland fire might be difficult to locate. Many of these systems are still in development, however, and false alarms are a concern.

---

[12]Infrared and ultraviolet technologies sense the electromagnetic radiation from a fire outside the visible band that humans can see. Temperature sensitive devices, such as heat sensitive resistant wires, do not sense radiation but react to temperature differentials.

[13]Sensors can also be placed around individual structures.

# Chapter 3

# Competing Concerns Affect Homeowners' Use of Protective Measures, but Efforts to Increase Their Use Are Under Way

Homeowners may not take steps to protect their homes from wildland fires because of the time or expense involved, competing concerns such as aesthetics or privacy, lack of understanding of the nature of wildland fire risks, and failure to recognize that they share responsibility for protecting their homes. Government agencies and other organizations are engaged in a variety of efforts to increase the use of protective measures, such as defensible space and fire-resistant building materials and design. These efforts include education to increase awareness by homeowners and others about steps they can take to reduce risks from wildland fire, monetary assistance to create defensible space, and laws requiring the use of protective measures. In addition, some insurance companies direct homeowners in high-risk areas to create defensible space. Fire officials told us that each of these approaches provided benefits but also posed challenges.

## Time, Expense, and Other Competing Concerns Affect Whether Homeowners Use Protective Measures

Time or the expense involved is one of the primary reasons behind homeowners' resistance to creating defensible space or installing fire-resistant roofs, fire officials told us.[1] Homeowners surveyed in three communities recently threatened by wildland fires in Colorado and Oregon also most frequently cited expense and time as impediments to creating defensible space.[2] Creating and maintaining defensible space involves trade-offs between money and time. Out-of-pocket expenses may be negligible when homeowners create defensible space themselves but completing the work can require substantial time and effort. Homeowners may also find it difficult to clear and transport any vegetation to appropriate disposal sites. Alternatively, homeowners can pay someone to create defensible space on their property. Fire officials estimate that the price of this work—including thinning trees and some replanting but not major landscaping—can be several thousand dollars or more depending on vegetation type and the topography of, and access to, a particular property. The New Mexico Forestry Division, for example, has estimated the price of creating 1 acre of defensible space around a structure in heavily forested areas in that state at about $1,700 to $2,400, although this estimate excludes the expense of removing large trees that are close to structures. A state

---

[1]Fire officials and representatives of the homebuilding industry said that concerns about cost can also affect homebuilders' decisions about building materials and landscaping.

[2]Holly Bender, Ingrid M. Martin, and Carol Raish, *What Motivates Homeowners to Protect Themselves from Risks?* (Boulder, Colo.: Integrated Resource Solutions, 2005).

forestry official estimated that removing such trees could cost $800 to $2,000 each. Second, regarding fire-resistant roofs, if homeowners wait until their existing roofs need replacement, cost does not have to be a major factor because fire-resistant roof-covering materials are available at similar cost to more flammable ones.

Homeowners may also be reluctant to create defensible space because of the importance they place on other considerations, such as the role of vegetation in their property's appearance, privacy, and wildlife habitat. Homeowners' concerns about the effect of defensible space on these features can be critical since such features influence homeowners' decisions to move nearer to wildlands in the first place. The design of defensible space is flexible, however, and can be done in ways that minimize the impact on appearance or wildlife habitat or even enhance them. When deciding whether to create defensible space, homeowners may also weigh the effects of landscaping on shade, energy efficiency, and water use, and they may sometimes receive contradictory advice from different government agencies about landscaping choices. For instance, water management districts in Florida promote landscaping choices that conserve water, but some of these choices may increase risk from wildland fire.

Another reason homeowners may not take protective measures is that they may not understand how wildland fires damage or destroy homes or how effective protective measures can be. An expert at the symposium convened for us by the National Academy of Sciences (NAS) said that because many homeowners think of wildland fires as intense crown fires, they do not believe that relatively simple steps like creating defensible space can be effective and, therefore, do not take such steps. On the contrary, however, defensible space can lessen the intensity of crown fires and, together with fire-resistant roofs and vents, can effectively protect against firebrands or low-intensity surface fires, which often damage structures. Forest Service researchers have reported that some homeowners do not think it worthwhile to create defensible space because they have seen a fire jump a six-lane highway. Fire officials said that these homeowners do not understand that defensible space is not intended to stop a fire from spreading but only to prevent it from reaching and igniting structures.

In addition, homeowners may not use protective measures because they believe that fire officials are responsible for protecting their homes and do not recognize they share in this responsibility. Fire officials told us that homeowners who have recently moved to the wildland-urban interface may not have experienced a wildland fire and may not realize their homes are at risk and that they should consider protective steps. Fire officials also said such newcomers may expect the same level of service they received in more urban areas and do not understand that rural areas may have fewer available firefighters and longer response times. Also, when a wildland fire burns near communities, so many houses may be threatened simultaneously that firefighters may be unable to protect them all. In such cases, defensible space and fire-resistant building materials greatly reduce a structure's risk.

## Education Helps Increase Awareness of Steps Homeowners and Others Can Take

Educating homeowners about the risks posed by wildland fire and the steps that can be taken to mitigate these risks is a critical step in increasing the use of measures to protect homes from wildland fires. Educating homeowners is effective in part because it can help overcome their reluctance to use protective measures, for instance, by showing them that defensible space can preserve or enhance their property's appearance and that even large trees can remain close to a structure, as long as defensible space is designed to protect those trees. Education also helps state and local government officials and professionals, such as landscape architects and planners, who influence where and how development occurs.

Federal, state, and local government agencies; universities and extension programs; nongovernmental organizations; and industry organizations are all involved in efforts to educate the public about protecting structures from wildland fires. The primary national effort to educate homeowners about protecting structures from wildland fire is the Firewise Communities program, which also promotes steps that state and local officials can take to educate homeowners. (The Firewise Communities Web site address, along with information on related Web sites, is included in app. IV.)[3] Because it seeks to increase voluntary use of protective measures, the Firewise Communities program requires homeowner and community involvement to be successful. To this end, since 1998, the Firewise

---

[3]Firewise Communities is jointly sponsored by the International Association of Fire Chiefs, National Emergency Management Association, National Association of State Fire Marshals, National Association of State Foresters, National Fire Protection Association, Federal Emergency Management Agency, U.S. Fire Administration, Forest Service, Bureau of Indian Affairs, Bureau of Land Management, Fish and Wildlife Service, and the National Park Service. Numerous state and local fire and forestry officials also participate in Firewise program activities.

Communities program has conducted more than 30 workshops, attended by approximately 3,000 people from 44 states, and has supported over 500 local or regional workshops reaching over 15,000 participants. The program has also distributed videos, books, brochures, and other materials that promote Firewise landscaping and construction. Finally, the program has recognized more than 100 communities in 26 states as "Firewise" communities. Homeowners in these communities, along with fire officials, assessed the community's wildland fire risk, developed a plan to mitigate those risks, and undertook activities to implement the plan.

Other education efforts are directed at state and local government officials and professionals, such as landscape architects and planners. For example, the American Planning Association and the National Fire Protection Association reported in February 2005[4] on approaches to educating planners about the risks wildland fires pose to communities and steps that local governments can take to reduce those risks. The report provides examples of planning approaches that have been adopted and discusses their shortcomings and is expected to be distributed to approximately 1,300 planning agencies nationwide. An American Planning Association official said that, as more development occurs in the wildland-urban interface, local governments must plan development wisely to help reduce the risk from wildland fire.

Examples of other education efforts from the states we visited include the following:

- The Institute of Business and Home Safety; the U.S. Forest Service; Alachua County, Florida; and others sponsored a demonstration project near Gainesville, Florida, that included landscaping a house to create defensible space and replacing the roof and siding with fire-resistant materials (see fig. 14). This project was intended to increase fire awareness among homeowners in the community and to show that creating defensible space could also be attractive and provide other amenities. Information on the project, including many photographs, was included on a Forest Service Web site so that other homeowners could view the project.[5]

---

[4]James Schwab, Stuart Meck, and Jamie Simone, *Planning for Wildfires* (Washington, D.C.: American Planning Association, 2005).

[5]See: http://www.interfacesouth.org/fire/firewisehome/. For additional information on this project, see http://www.firewise.org/vrhome/.

**Figure 14: Before and After Photos of a Firewise Demonstration Home**

Before                    After

Before                    After

Source: Larry Korhnak.

- The Sonoran Institute and the National Association of Counties sponsored a September 2004 workshop attended by county officials from Idaho, Montana, and Wyoming to discuss the role of zoning and other growth management approaches in reducing the wildland fire risk to new development. The workshop discussed the costs associated with new development in the wildland-urban interface, such as increased fire suppression costs, and the importance of land-use planning and other approaches to reduce risks from wildland fires, according to the workshop organizer.

- In Florida, the Department of Community Affairs and Division of Forestry published a handbook in April 2004 that describes different wildland fire mitigation strategies that communities in Florida have adopted. The handbook contains information directed at homeowners, homebuilders, government officials, and professionals such as planners and landscape architects.[6] The section on landscaping, for instance, provides examples of less flammable plants—such as azaleas, dogwoods, and oaks—appropriate for planting in areas at risk of wildland fire.

Federal, state, and local officials we met with said that although education efforts are critical to increasing awareness of the risks of wildland fire and of the steps that can be taken to reduce those risks, they face challenges that will take time to overcome. Because homeowners have concerns other than reducing the risk from wildland fires, providing information on risks and steps to reduce those risks, officials and researchers said, may not result in homeowners taking action. Similarly, providing information to state or local government officials—for instance, about laws or land-use planning strategies to reduce the risks to structures from wildland fire— may not lead those officials to adopt such measures. To increase the likelihood of success, symposium experts and other officials said those conducting education programs should recognize that multiple approaches exist to making a structure more fire-resistant, and educators should assist homeowners to find the approach that best suits their needs. Information describing defensible space, for instance, can show several different ways of making a structure more fire-resistant so that homeowners can see the effect on the appearance of their property.

---

[6]Florida Department of Community Affairs and Florida Department of Agriculture and Consumer Services, *Wildfire Mitigation in Florida: Land Use Planning Strategies and Development Practices* (Tallahassee: April 2004). (Available at http://www.dca.state.fl.us/fdcp/DCP/publications/Wildfire_Mitigation_in_FL.pdf).

## Financial and Other Assistance Encourages Homeowners and Communities to Take Action

Federal, state, and local agencies are also taking steps to directly assist individual homeowners and communities in creating defensible space and reducing hazardous fuels. This assistance can help homeowners balance the trade-offs between expense and time in creating defensible space.

Under the National Fire Plan,[7] federal firefighting agencies provide grants or otherwise assist in reducing fuels on private land. For instance, the Forest Service provided approximately $11.6 million (adjusted for inflation) to the New Mexico Forestry Division from fiscal year 2001 through 2004 that the state could use to assist reduction of fuels on nonfederal land.[8] Grants to reduce fuels on private property typically require the homeowner to pay a portion of project costs.[9] National Fire Plan funds have also been used to create fuel breaks around communities. For example, the Washington Department of Natural Resources received a $340,000 grant that it used to create a fuel break around the town of Roslyn, reducing fuels in an approximately 150-foot-wide buffer zone. Fire officials told us the fuel break by itself would not prevent a wildland fire from entering the community, but that it would assist suppression efforts by reducing fire intensity close to the community. The grant also funded creation of defensible space for an additional 144 homes located outside the fuel break.

State and local governments have provided similar assistance. The Florida Division of Forestry, for instance, has used state and federal funds to establish four mitigation teams that reduce fuels on private lands by conducting prescribed burns and mechanically removing vegetation to

---

[7]The National Fire Plan was developed by the Department of Agriculture and the Department of the Interior after severe wildland fires in 2000. In fiscal year 2001, Congress almost doubled funding for federal firefighting agencies to help meet the plan's objectives to (1) increase fire suppression preparedness; (2) rehabilitate and restore lands and communities damaged by wildland fire; (3) reduce hazardous fuels; and (4) assist communities through education, hazard mitigation, and training and equipment for rural and volunteer fire departments.

[8]The $11.3 million includes funds provided under the National Fire Plan and other federal programs. In addition to reducing fuels on nonfederal land, some of these funds may also have been used to assist local fire departments or to otherwise address wildland fire concerns.

[9]Grants to the East Mountain community near Albuquerque, for instance, capped eligible project costs at $1,700 for the acre immediately surrounding the house and approximately $1,150 per acre for up to 4 additional acres if they posed a risk to the structure; the homeowner was required to pay 30 percent of eligible costs.

create fuel breaks around communities at high risk of wildland fires. In other cases, local governments have helped homeowners to chip or remove vegetation produced by the creation of defensible space. Santa Fe County, New Mexico, for instance, bought two grinders in 2003 to chip vegetation and established locations where homeowners from participating communities could bring plant material they removed from their property. The county fire marshal told us that this program had assisted approximately 1,000 residents.

Federal, state, and local fire officials and homeowners told us that efforts such as these are helpful but also raise some concerns. First, because vegetation grows back, fuel breaks and defensible space need to be maintained to be effective (see fig. 15). To address this concern, Florida Division of Forestry officials told us that the division requires communities it assists to sign an agreement to maintain the defensible space or fuel breaks. Second, fire officials said it is difficult to identify sources for grants and other assistance. In some of the states we visited, federal and state officials are working to assist homeowners and local officials to identify such sources. Firewise Communities program officials said they have identified assistance available in many states and posted a list on their Web site (see app. IV). Finally, some homeowners raised concerns about grant eligibility requirements. New Mexico, for instance, requires grants or assistance to be distributed to homeowners through another government entity, for example, a city fire department or local governmental district. If a local government is not able to sponsor the grant, residents must incorporate as a not-for-profit organization to be eligible, a process a participating homeowner told us was frustrating and time-consuming.

**Figure 15: Fuel Break near Roslyn, Washington, Shown after Construction and 3 Years Later**

Source: Washington State Department of Natural Resources.

Source: GAO.

## State or Local Laws May Require Protective Measures

States, counties, and cities can adopt laws designed to reduce the risk to homes from wildland fires by requiring protective measures, such as creation of defensible space or the use of fire-resistant building materials.[10] Local governments can also improve fire safety through land-use planning, by restricting development or requiring additional protective measures in particularly fire-prone areas. Ventura County, California, fire officials attribute the relatively few houses in that county damaged by the 2003 Southern California fires to, in part, the county's adoption and enforcement of laws requiring 100 feet of defensible space and the use of fire-resistant building materials. Such steps are particularly effective at reducing the risk of wildland fires for new developments because it is cheaper to select building materials and incorporate fire-resistant community design before construction begins. After the 2003 Southern California fires, for instance, San Bernardino County officials reported that communities developed more recently under requirements regarding vegetation and building materials sustained far less damage during those fires than did older

---

[10]State or local governments can also adopt laws that establish standards for water supply and emergency access. These requirements assist suppression efforts and are beyond the scope of this study.

communities.[11] Symposium experts told us that as more people move into the wildland-urban interface, the benefits of local governments' requiring protective measures are likely to increase.

States or local governments can adopt or adapt model laws requiring protective measures developed by one of several organizations, including the International Code Council and the National Fire Protection Association, or they can develop their own requirements. Laws adopted by individual jurisdictions vary but can include requirements for the creation of defensible space and use of fire-resistant building materials and design (see table 1). Some jurisdictions have applied land-use planning to restrict development in areas that are at particularly high risk of wildland fire. Alachua County, Florida, for instance, amended its comprehensive plan in 2002 to address wildland fire risks. Under the plan, the county will not approve new developments unless they are designed to provide adequate protection from wildland fire, as determined by the county fire chief.

---

[11]Governor's Blue Ribbon Fire Commission, *Report to the Governor* (Sacramento, Calif.: 2004).

**Table 1: Examples of Laws Requiring Protective Measures Adopted by Jurisdictions in Five States GAO Visited**

| Jurisdiction | Requirements |
| --- | --- |
| **States** | |
| California | In 2005, California increased its statewide defensible space requirements from 30 feet to 100 feet and explicitly allowed local governments or insurance companies to require even greater clearance. In very-high-fire-hazard-severity areas, class A roofing materials are required for new construction. |
| Washington | In 1999, the state's Department of Natural Resources developed a model ordinance recommending that structures in areas at risk from wildland fire maintain a minimum of 50 feet of defensible space and use fire-resistant building materials, among others things. Although not binding, state officials disseminated the model ordinance to county and city officials. |
| **Counties** | |
| Ada County, Idaho | The county has identified lands at high risk of wildland fire and, since 1997, has required homeowners in this area to maintain at least 50 feet of defensible space around new structures. New construction in the high-risk area must comply with additional requirements, including at least class B roofing materials; screened vents; enclosed eaves; nonflammable gutters; and fire-resistant exterior walls, windows, and decks. |
| Ventura County, California | The county requires 100 feet of defensible space and further recommends that owners of homes at particularly high risk increase defensible space to 200 feet. In high-fire-hazard areas, the county requires structures be constructed with class A roofing materials and fire-resistant building materials. In addition, all new structures larger than 5,000 square feet or more than 5 miles from a fire station are required to install a sprinkler system. |
| **Cities** | |
| Ormond Beach, Florida | Since 2003, new construction in areas identified by the city as at medium or high risk for wildland fires must develop vegetation management plans establishing at least 30 feet of defensible space around a structure. A 30-foot buffer zone must also be created around the perimeter of a new planned development or residential subdivision and be maintained by homeowners or a homeowners' association according to a management plan approved by the city. |
| Santa Fe, New Mexico | In 2004, fire officials worked with city officials to modify a city ordinance requiring homes built on ridgelines or in the foothills to plant and maintain evergreen trees at the same density as in the adjacent natural landscape to reduce the visual impact of such development. Under the amended ordinance, homeowners may use some deciduous trees, which are less flammable, and can also reduce vegetation density to a level approved by the city. |

Source: GAO analysis of state, county, and city data.

For laws and land-use planning to be an effective tool in reducing damage to structures from wildland fires, individual state and local governments must adopt and enforce them. State and local fire officials told us that although no one has compiled a complete list of governments that have adopted laws designed to reduce the risk to structures from wildland fire, many at-risk jurisdictions have adopted laws, and many others have not.[12]

---

[12]The Forest Service's Southern Research Station has compiled a list of state and local governments reporting they have adopted codes or other measures designed to reduce the risk to structures from wildland fires. This information is available on the World Wide Web at www.wildfireprograms.com/.

Symposium experts and fire officials said that the primary reason for not adopting laws is community opposition to them. Other officials, homeowners, and a homebuilding industry representative expressed concern that some proposed laws may not offer significant additional protection from wildland fire or may not be cost-effective, considering the low probability that a home would be destroyed. Symposium experts recognized opposition to such laws but stressed the importance of state and local governments' adoption of them. Moreover, once adopted, laws must be enforced to be effective. Effective enforcement requires confirming that homeowners and others comply with requirements and ensuring that requirements are not weakened by exemptions for individual developments. Ventura County officials told us that active enforcement of their laws was an important factor in the relatively few houses damaged in that county during the 2003 Southern California fires.[13] They also said that compliance increased as homeowners became more familiar with the requirements and the enforcement program. Nevertheless, symposium experts said many fire departments, counties, and cities do not have sufficient resources to effectively enforce laws, or they may be pressured by homeowners or developers not to. In addition, the effectiveness of laws can be undercut by variances exempting individual developments from specific requirements, such as emergency access. In some cases, officials said such variances may be warranted, for instance if the proposed development is not at significant risk, or if additional measures are incorporated to increase protection. In other cases, county or city officials may be pressured to approve a variance even if the development is at risk.

## Some Insurance Companies Direct Homeowners to Use Protective Measures

Although wildland fire has not resulted in significant losses for the insurance industry in comparison with other disasters, some insurance companies have instituted programs designed to increase policyholders' use of protective measures in some at-risk areas. Since 1993, for instance, one major company has evaluated high-risk properties in California for defensible space before underwriting new policies. A company official said that 200 to 500 feet of defensible space is often required, depending on factors such as topography, vegetation density, and type of construction. In 2004, the company began expanding this program to other western states. Another major company initiated a pilot program in 2003 in Colorado, Utah,

---

[13]In April each year, the county fire department notifies approximately 14,000 homeowners that they need to create defensible space by June 1. If a homeowner does not do so, the county charges him or her for the cost of a contractor to do the work.

and Wyoming, under which the company inspected properties of policyholders living in certain high-risk areas in those states and notified policyholders of any actions needed to establish defensible space according to the standards required or recommended by their local fire departments. Policyholders would have at least 18 months to perform any work needed to meet those standards, according to the company official in charge of the program and, if the corrective actions were not completed, the company could choose not to renew the policy. The official said that it is too early to evaluate the program's success but he expects the program to continue and perhaps expand to other regions of the country.

Some fire officials have said that the insurance industry should take a larger role in encouraging use of protective measures, such as by offering discounts on premiums to policyholders who have defensible space. Insurance industry officials we spoke with said that the share of premiums associated with wildland fire risk is relatively low and would not provide a meaningful incentive for homeowners. Although industry losses have been low historically, officials from the Insurance Services Office told us that recent trends toward increased fire severity and more people living in at-risk areas mean that future losses may be higher.

## Possible Federal Government Actions to Increase Use of Protective Measures

As we previously mentioned, homeowners and state and local governments have the primary responsibility for taking preventive steps to protect homes from wildland fires. Nevertheless, the federal government currently funds education for homeowners and communities, primarily through the Firewise Communities program, and provides grants to states and communities to use on preventive measures to protect structures, under the National Fire Plan and other sources. Key to choosing the appropriate approach will be determining what the federal role should be in this area, given that the majority of the structures damaged by wildland fires are located on private property, and losses are normally covered by the fire portion of homeowners' insurance. In addition, although many homes are at risk from wildland fire, only a small fraction of those are actually damaged or destroyed in any given year, and damages and insured losses from wildland fire are significantly less than from either other natural disasters or other types of structure fires.

Should the federal government choose to continue or change its role, it can use a variety of policy options to motivate or mandate homeowners to implement measures to protect structures from wildland fires. These options include education partnerships, grants to states and localities to promote the use of protective measures, tax incentives, and building and land use regulations.[14] However, additional information in several areas would be helpful in more clearly defining the problem and determining the appropriate level of federal efforts to address it. Such information includes the scope and scale of the risk to homes from wildland fires, the actual losses incurred from wildland fires, the extent of efforts homeowners are already making to address wildland fire risks, and the extent to which homeowners cannot obtain private insurance. Most of this information, including the scope and scale of the risk, is not readily available or easily quantifiable.

There are three main considerations regarding education partnerships and grants to undertake preventive measures. First, because resources are scarce, spending decisions must be based on a careful assessment of whether the benefits to the nation from these efforts to reduce the risk to privately owned structures exceed their costs. Second, it is important to strike a balance between accountability and flexibility. Accountability can be achieved by establishing performance measures and outcome goals and measuring results. Doing so would allow flexibility in how funds are used, while at the same time ensuring national oversight. For example, information measuring the results and the effectiveness of federal grant making under the National Fire Plan would be useful in determining whether continued or additional funding for the program is needed. However, developing the appropriate performance measures is complicated because it is difficult to determine the number of structures that would have been destroyed or damaged if preventive measures had not been taken. The third consideration is targeting the funds to those with the greatest need. To effectively target grants to address the greatest threats to structures from wildland fires requires information on the relative risks from wildland fires faced by different communities.

Tax incentives are the result of special exclusions, exemptions, deductions, credits, deferrals, or tax rates in the federal tax laws. Unlike grants, tax incentives do not generally permit the same degree of federal targeting and oversight, and they generally are available to all potential beneficiaries who satisfy congressionally established criteria. In the case of wildland fire, while potentially millions of homes are at risk and might qualify for tax incentives, the number of homes that actually are damaged or destroyed by

---

[14]Some of these options can be carried out under existing law; others would require new legislation.

wildland fires each year is a small fraction of those at risk. To make a reasoned judgment about the effectiveness of this policy option, additional information would be needed on the number of homeowners that could qualify for tax incentives and possible cost and benefits of the incentives.

The federal government has little authority over land-use planning or building on private land. The authority to develop, adopt, administer, and enforce building and land-use regulations has traditionally rested with the states, which in turn have delegated some or all of their authority to local governments. In a few instances, such as the Coastal Zone Management Act, the federal government has provided incentives for state and local governments to adopt development plans that meet specific criteria. Congress could provide similar incentives for state and local governments to adopt building and land-use regulations addressing threats to structures from wildland fires. However, state and local officials we spoke with expressed concern about having the federal government take a role in these types of regulations rather than leaving responsibility at the state and local level.

# Chapter 4

# Effective Adoption of Technologies to Achieve Communications Interoperability Requires Better Planning and Coordination

While a variety of existing technologies can help link incompatible communications systems and others are being developed to provide enhanced interoperability, effective adoption of any technology requires planning and coordination among federal, state, local, and tribal agencies that work together to respond to emergencies, including wildland fires. Without such planning and coordination, new investments in communications equipment or infrastructure may not improve the effectiveness of communications between agencies. The Department of Homeland Security (DHS) is leading federal efforts to address interoperability problems across all levels of government, but as we previously reported, progress so far has been limited. Some state and local government efforts are also under way to improve communications interoperability.

## Technologies Can Enhance Communications Interoperability

A number of current and emerging technologies can help overcome differences in frequencies or communications equipment and improve communications interoperability among firefighting agencies. These include technologies for short-term solutions—often called patchwork interoperability—to interconnect disparate communications systems and longer-term improvements to communications equipment and infrastructure.[1]

### Patchwork Interoperability

Patchwork interoperability uses technology to interconnect two or more disparate radio systems so that voice or data from one system can be made available to all systems. The principal advantage of this solution is that agencies can continue to use existing communications systems, an important consideration when funds to buy new equipment are limited. According to an official from DHS's Office for Interoperability and Compatibility, a major disadvantage to all patchwork solutions is that they require twice as much spectrum since they have to tie up channels on both connected systems. Three main patchwork technologies are currently

---

[1]One solution to improve interoperability is to have a cache of portable radios that can be distributed to responding personnel during an emergency. For example, Florida has a system of radio caches, one cache located in each of the seven regions of the state. The nation's cache of approximately 8,000 radios is operated by the National Interagency Incident Communications Division at the National Interagency Fire Center in Boise, Idaho. These radios are routinely used for large fires and also for other incidents including hurricanes and the terrorist attacks on September 11, 2001, according to a National Interagency Fire Center official.

available. Appendix V provides more detail about each of these technologies.

- *Audio switches* provide interoperability by connecting radio and other communications systems to a device that sends the audio signal from one agency's radio to all other connected radio systems. Audio switches can interconnect several different radio systems, regardless of the frequency bands or type of equipment used.

- *Crossband repeaters* provide interoperability between systems operating on different radio frequency bands by changing frequencies between the two radio systems.

- *Console-to-console patches* link the dispatch consoles of two radio systems so that the radios connected to each system can communicate with one another. Dispatch consoles are located at the dispatch center where dispatchers receive incoming radio calls.

Audio switches are easily transportable and can be used to create temporary interoperability, which makes them useful for wildland firefighting where multiple agencies temporarily come together to fight the fire. In addition to ease of transport, audio switches are flexible and allow a variety of communications systems, including radio and telephone, to be connected. Public safety agencies in several localities, including Washington, use them. In addition, the National Interagency Incident Communications Division at the National Interagency Fire Center (NIFC) recently purchased two of these devices to use to connect radio systems during major public safety incidents. An audio switch costs about $7,000[2] without the radio interface modules[3] or cables. Each interface module costs about $1,100, and cables are available for about $140 each.

A crossband repeater provides interoperability between systems operating on different radio frequency bands by changing the frequency of the signal received and sending it out on another frequency. For example, a

---

[2]Cost estimates for communications technologies were obtained from the General Services Administration (www.gsaadvantage.gov) or directly from manufacturers.

[3]A radio interface module is a device that plugs into the chassis of the audio switch. Each radio system being interconnected through the switch connects through a radio interface module. The interface module separates out the audio and other signals needed to control the radios connected to the switch.

crossband repeater can receive a VHF (very high frequency) signal and retransmit it as a UHF (ultrahigh frequency) signal. Crossband repeaters can connect base stations[4] or mobile radios, whether hand carried or in vehicles. A variety of crossband repeaters are available ranging in price from $4,000 to $33,000 each. Crossband repeaters can cost more than audio switches, which may put them beyond the reach of jurisdictions with limited funding. Still, according to a communications specialist at NIFC, crossband repeaters are an effective interoperability solution often used by federal firefighting agencies.

Unlike audio switches or crossband repeaters, a console-to-console patch is not an "on-the-scene" device but instead the connection occurs between consoles located at the dispatch centers where calls for assistance are received. The costs of such a connection vary widely, depending on whether consoles are patched together temporarily over a public telephone line, or permanently over a dedicated leased line or a dedicated microwave or fiber link.[5] The costs for a dedicated leased line would consist primarily of recurring telephone line charges. In contrast, a microwave link connecting two locations about 15 to 25 miles apart could require an initial investment of about $70,000.

## Improved Communication Systems

Other interoperability solutions involve developing and adopting more sophisticated radio systems that follow common standards or can be programmed to work on any frequency and to use any desired modulation type, such as AM or FM. Project 25 radios, software-defined radios, and Voice over Internet Protocol are the primary examples of these improved communications systems. Appendix V provides more detail about each of these technologies.

- *Project 25 radios*, which are currently available, must meet a set of standards for digital two-way radio systems that allow for interoperability between all jurisdictions using these systems.

---

[4]A base station contains the equipment for transmitting and receiving the radio signals that allow portable radios to communicate with each other.

[5]A leased line refers to a permanent telephone connection set up by a telecommunications provider between two geographic locations. A fiber link uses light sent over a glass or plastic fiber to carry communication signals. A microwave link uses radio beams of extremely high frequencies to send information between two fixed geographic sites.

- *Software-defined radios*, which are still being developed, are designed to transmit and receive over a wide range of frequencies and use any desired modulations, such as AM or FM.

- *Voice over Internet Protocol* treats both voice and data as digital information and enables their movement over any existing Internet Protocol data network.[6]

Project 25, also called APCO 25, was established in 1989 to provide detailed standards for digital two-way wireless communications systems so that all purchasers of Project 25-compatible equipment can communicate with each other.[7] They can also communicate with older, analog radios. Project 25 radios, at about $1,700 to $2,500 each, cost more than other available radios that cost around $1,200 each. Federal, state, and local officials we spoke with agreed that, while Project 25 radios could provide interoperability benefits, funding and other limitations will likely result in phased adoption. For example, a federal communications specialist said that the Forest Service will be purchasing Project 25 radios over a 10-year replacement cycle. As of December 2003, the state of Washington had about 400 Project 25-compatible radios, of a total of 8,000 portable radios owned by the state. None of the 400, however, are owned by the agency responsible for wildland firefighting.

Software-defined radios and Voice over Internet Protocol appear to hold promise for improving interoperability among firefighting and other public safety agencies. Voice over Internet Protocol offers the flexibility to transmit both voice and data over a data network. This could be useful for firefighting agencies that need weather and other information when making decisions affecting fire suppression efforts. However, no standards exist for radio communications using Voice over Internet Protocol and, as a result, manufacturers have produced proprietary systems that may not be interoperable. Software-defined radios will allow interoperability among agencies using different frequency bands, different operational modes (digital or analog), proprietary systems from different manufacturers, or

---

[6]In some cases, this is the Internet; and in others, it is a private data network.

[7]Project 25 standards are being developed jointly by the Association of Public Safety Communications Officials International; the National Association of State Telecommunications Directors; the National Telecommunications and Information Administration; the Department of Homeland Security's National Communications System; and the Department of Defense.

different modulations (such as AM or FM). However, software-defined radios are still being developed and are not yet available for use by public safety agencies.

# Planning and Coordination Are Key to Improving Communications Interoperability

In the past, public safety agencies have depended on their own stand-alone communications systems, without considering interoperability with other agencies. Yet as firefighting and other public safety agencies increasingly work together to respond to emergencies, including wildland fires, personnel from different agencies need to be able to communicate with one another. Reports by GAO,[8] the National Task Force on Interoperability, and others have identified lack of planning and coordination as key reasons for lack of communications interoperability among responding agencies. According to these reports, federal, state, and local government agencies have not worked together to identify their communications needs and develop a coordinated plan to meet them.

Whether the solution is a short-term patchwork approach or a long-term communications upgrade, officials we spoke with explained that planning and coordination among agencies are critical for successfully determining which technology to adopt and for agreeing on funding sources, timing, training, maintenance, and other key operational and management issues. States and local governments play an important role in developing and implementing plans for interoperable communications because they own most of the physical infrastructure for public safety systems, such as radios, base stations, repeaters, and other equipment.

In recent years, the federal government has focused increased attention on improving planning and coordination to achieve communications interoperability. The Wireless Public Safety Interoperable Communications Program (SAFECOM) within DHS's Office of Interoperability and Compatibility[9] is responsible for addressing interoperability and compatibility of emergency responder equipment, including communications. SAFECOM was established to address public safety communications issues within the federal government and to help state,

---

[8]See GAO, *Homeland Security: Challenges in Achieving Interoperable Communications for First Responders*, GAO-04-231T (Washington, D.C.: Nov. 6, 2003).

[9]The Wireless Public Safety Interoperable Communications Program, otherwise known as SAFECOM, was first established as an Office of Management and Budget e-initiative in 2001.

local, and tribal public safety agencies improve their responses through more effective and efficient interoperable wireless communications. We reported, in April 2004, that SAFECOM had made limited progress in addressing its overall program objective of achieving communications interoperability among entities at all levels of government.[10] Further, we reported in July 2004 that the nationwide data needed to compare current communications interoperability conditions and needs, develop plans for improvement, and measure progress over time were not available. In that report, we recommended, among other things, that DHS continue to develop a nationwide database and common terminology for public safety interoperability communications channels and assess interoperability in specific locations against defined requirements. DHS agreed with these recommendations.

DHS has been working on a number of initiatives since SAFECOM began. In March 2004, SAFECOM published a *Statement of Requirements for Public Safety Wireless Communications and Interoperability* to begin identifying the specific communications needs of public safety agencies. The statement of requirements is being updated to further refine the information and is scheduled for release to the public by June 30, 2005. In addition, SAFECOM published the *Statewide Communication Interoperability Planning Methodology* in November 2004, which was developed in a joint project with the commonwealth of Virginia. The methodology describes a step-by-step process for developing a locally driven statewide strategic plan for enhancing communications interoperability, including key steps and time frames. Finally, in January 2005, SAFECOM awarded a contract to develop and execute a nationwide interoperability baseline study, which SAFECOM officials anticipate will be completed by December 30, 2005. According to officials, this study will provide an understanding of the current state of interoperability nationwide, as well as serving as a tool to measure future improvements made through local, state, and federal public safety communications initiatives.

In addition to federal efforts, a variety of steps have been taken by state and local agencies. Several states, including California, Florida, Idaho, Missouri, and Washington, as well as the commonwealth of Virginia have developed statewide groups to address communications interoperability. For example, Washington established the Washington State Interoperability

---

[10]See GAO-04-494.

Executive Committee in July 2003. According to a state official, the committee was created to ensure communications interoperability by managing and coordinating the state's investments in communications systems. The committee's responsibilities included completing an inventory of state government-operated public safety communications systems, preparing a statewide public safety communications plan, establishing standards for radios, seeking funding for wireless communications, and fostering cooperation among emergency response organizations. By December 2003, the group had developed an inventory of state-operated public safety communications systems and in March 2004 the group published an interim statewide public safety communications systems plan.

In some cases, neighboring jurisdictions or public safety agencies are working together to address communications issues. To improve interoperability between federal, state, and local responders in Los Angeles County, the Los Angeles Regional Tactical Communications Systems Executive Committee was formed. According to a county fire official, barriers to interoperability in the county and with neighboring counties include agencies operating on different radio frequencies and using incompatible technologies, as well as a lack of funding for communications systems. The group is using a two-track effort to improve communications: (1) acquiring and using interconnection devices, such as audio switches, with existing communication resources to enhance interoperability and (2) rebuilding communications infrastructures for improved interoperability in the long-term. As of February 2005, the Los Angeles County Fire Department had acquired three audio switch units, according to a county fire official.

# Use of Military Assets to Fight Wildland Fires

The federal government and the states can provide a variety of military assets, including aircraft and military personnel, to assist in wildland firefighting. The process used to request, authorize, and deploy these assets varies depending on whether the asset is under federal or state control. The National Interagency Coordination Center (NICC), which coordinates firefighting resources on a national level, is responsible for requesting federal military aid for firefighting from the Department of Defense (DOD). A state firefighting agency is responsible for requesting state military aid from its governor's office. Federally controlled military resources are normally used only after the nation's federal, state, local, tribal, and contract firefighting resources have been depleted. Various laws, agreements, and policies specify when federal military assets can be used and the process for requesting them. According to key participants in the process, current procedures for requesting and using federal military resources to fight wildland fires have generally worked well and continue to be appropriate. Federal military resources have been used to fight wildland fires in 9 out of the 16 years from 1988 through 2003.

## Types of Military Assets Available for Firefighting

The federal government and the states can provide a variety of military equipment and personnel to assist in firefighting, including large fixed-wing aircraft that can be converted to tankers for dropping retardant on fires; helicopters to carry personnel, equipment, or external buckets to drop water on fires; battalions of military personnel to serve as firefighters or mop-up crews; or other specialized personnel and equipment.[1] The federal government controls active military, military reserve, and federalized

---

[1]DOD military bases can also enter into mutual aid agreements with federal, state, or local firefighting agencies. Depending on the terms of these agreements, civilian firefighting forces stationed at a military base can either provide or receive assistance. It was beyond our scope to gather representative data on how extensively such military assistance is actually used for firefighting in wildlands or the wildland-urban interface. Consequently, we excluded such assistance from our discussion.

National Guard assets,[2] and state governments control all other National Guard assets.[3]

One of the primary military aids for wildland firefighting is the Modular Airborne Fire-Fighting System (MAFFS). This joint program of the Forest Service and DOD has been operating since 1974. When contracted air resources[4] are not readily available,[5] the Forest Service can request C-130 fixed-wing aircraft from DOD. There are eight of these aircraft in the nation. Six are under the control of state National Guard units: two each in California, North Carolina, and Wyoming. The remaining two are under the control of the Air Force Reserve in Colorado.[6] The Forest Service owns self-contained, reusable 3,000-gallon aerial fluid dispersal systems, which can be installed on these aircraft for holding fire retardant until it is dropped on a wildland fire (see fig. 16).

---

[2]The National Guard has both a federal and a state mission. The federal mission is to be available for prompt mobilization during war and provide assistance during national emergencies, such as natural disasters or civil disturbances. When not mobilized or under federal control, National Guard units report to the governors of their respective states or territories.

[3]Local military commanders or responsible officials of DOD agencies may, under the "immediate response criteria," take necessary action to save lives, prevent human suffering, or mitigate great property damage prior to receiving approval to do so.

[4]According to officials from the National Interagency Fire Center, the Forest Service and the Department of the Interior have a fleet of approximately 700 aircraft, including both large and small fixed-wing aircraft and helicopters. Many of these are contracted aircraft. Until May 10, 2004, there were also 33 privately owned large air tankers under contract to the Forest Service, which were used to drop retardant on wildland fires. These contracts were cancelled, however, due to concerns about the safety and airworthiness of these aircraft. According to an NIFC official, a contract was issued in March 2005 for at least 20 large air tankers, pending operational service life determination.

[5]The agreements with the states of California and Wyoming, in effect, define "readily available" as able to be moved into the local area within 2 hours.

[6]The Governors of California, North Carolina, and Wyoming may also activate the Air National Guard Unit in their state for MAFFS missions within state boundaries provided such action is covered by an appropriate Memorandum(s) of Understanding and Collection Agreement with the military authority and the Forest Service. They must request permission to use the Forest Service-owned equipment.

**Figure 16: MAFFS Used for Wildland Firefighting**

Source: DOD.

A variety of helicopters are available to transport personnel, supplies, or equipment, or they can be outfitted with external water buckets to drop water on fires (see fig. 17). For example, a UH-1 helicopter can carry 420 gallons of water, and a Chinook 47 can carry 2,600 gallons.

**Figure 17: A Helicopter Using a Water Bucket**

Source: DOD.

Other military assets may also assist in firefighting. These can include military personnel for firefighting or for mop-up activities ensuring that the fire has been completely extinguished after the main fire suppression effort. The military may also provide equipment or personnel specializing in communications, geospatial imagery, remote weather forecasting, or medical services.

## Process for Requesting and Mobilizing Military Assets for Firefighting

To begin the process of requesting federal military aid, NICC, located at the National Interagency Fire Center (NIFC) in Boise, Idaho, must first determine if such aid is needed. NICC is responsible for monitoring fire activity and firefighting resource availability across the nation. On the basis of this information, the NICC coordinator recommends a national preparedness level ranging from 1 to 5. Preparedness level 1 indicates minimal fire activity nationwide with little or no commitment of national resources. In contrast, preparedness level 5 indicates that several geographic areas[7] are experiencing major incidents having the potential to exhaust all agency fire resources. As the nation moves to level 3 or 4, the NICC coordinator advises DOD that a defense coordinating officer (also called a military liaison officer) is needed to assist NIFC in working with the military, helping with terminology, and coordinating with DOD organizations in case military assets are needed to assist in firefighting. If level 5 is reached and additional firefighting resources are needed, NIFC may request assistance from DOD.[8] Because wildland firefighting is not the primary mission of DOD, federally controlled military resources are normally used only after the nation's federal, state, local, tribal, and contract firefighting resources have been depleted. If DOD officials believe that the request meets the criteria laid out in DOD Directive 3025.15, which includes legality, appropriateness, and cost criteria, they may make resources available (see fig. 18).

---

[7]To provide cost-effective and timely coordination of emergency response, the nation is divided into 11 geographic areas, each of which is served by a geographic area coordination center.

[8]MAFFS may be requested when contracted air resources are not readily available, which is not directly related to the nation's preparedness level.

**Figure 18: Process for Requesting Military Assistance**

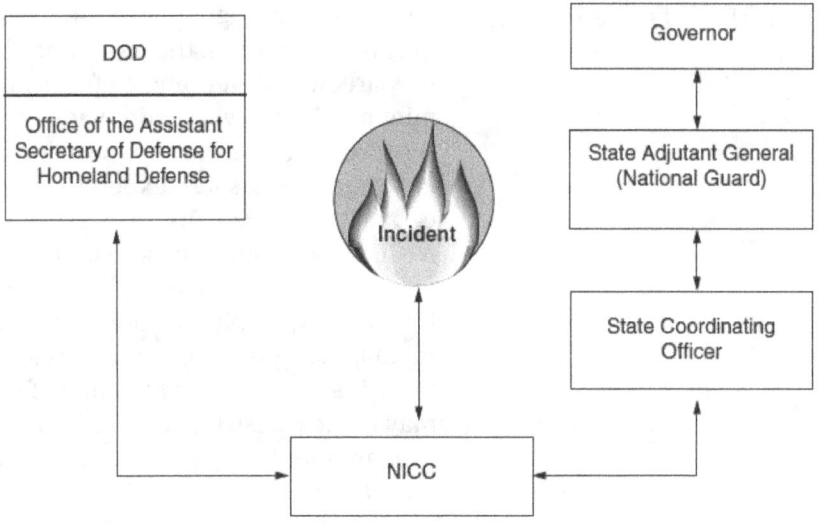

Federal military assistance

- Can request military assistance if civilian resources are unavailable

- DOD evaluates request against criteria including:

  --Availability of resources

  --Impact on ability to perform primary mission

State National Guard

- Agreement must be in place

- Governor signs proclamation of emergency

Sources: GAO and DOD.

To request state military aid, the state agency responsible for wildland firefighting coordinates with the governor's office, which controls these National Guard assets. State-controlled assets are normally used only after the governor has declared a state of emergency.

Advance planning is needed to facilitate the mobilization of military assets for firefighting. NIFC policies and procedures state that qualifying a military unit for a nondesignated military mission, such as dropping water on a wildland fire, is a major undertaking, requiring extensive planning by both the military and the firefighting agencies. For example, a number of steps must be taken before a MAFFS crew and aircraft are ready for a wildland firefighting mission. Before the fire season starts, initial and refresher training is required for pilots and flight crews. When a firefighting agency requests MAFFS assistance, qualified military aviation units must be identified, approved, and recalled to the base, which may take several hours. In addition, the plane must be readied for firefighting by removing external fuel tanks, loading and testing the MAFFS, conducting preflight checks, and fueling the aircraft. Together these steps may require as long as 24 hours to complete.

For federally controlled military personnel, approximately 5 days are needed for training, supplying them necessary clothing and equipment, and traveling to the fire. These personnel need classroom and actual firefighter training to learn about fire behavior; firefighting equipment and techniques; and the proper use of safety equipment, such as fire shelters. Carrying out these activities is a major undertaking, given that typically a battalion— consisting of 25 crews of 20 persons each, or more than 500 individuals including all supporting personnel—is typically mobilized for firefighting. Up to 60 federal firefighting managers and other personnel are needed to train and supervise military personnel, according to federal officials, which can be difficult in severe fire seasons when there are often not enough personnel to fill all demands. Federal firefighting and DOD officials explained that it would not be an effective use of resources to train military personnel ahead of the fire season because it is uncertain whether military assistance will be needed in any given fire season, and any personnel trained may be deployed for other missions and unavailable when called upon for wildland firefighting.

## Laws, Agreements, and Policies Governing the Use of Military Assets under Federal and State Control

The primary law that allows federally controlled military assistance in wildland firefighting is the Economy Act of 1932,[9] which provides general authority to federal agencies to use the services of other agencies. The act authorizes an agency to obtain the services of another agency when

- funds are available,

- the head of the ordering agency decides it is in the best interest of the government,

- the performing agency is able to provide or obtain by contract the ordered good or services, and

---

[9]31 U.S.C. § 1535.

- the head of the ordering agency decides that the resources cannot be provided by contract "as conveniently or cheaply by a commercial enterprise."

As the Office of Management and Budget (OMB) pointed out in its 2004 report,[10] the Economy Act requires that officials evaluate whether the needed goods or services can be provided as conveniently or cheaply by a commercial enterprise, but it does not require that all commercial resources be exhausted before requesting assistance from another federal agency.

The Robert T. Stafford Disaster Relief and Emergency Assistance Act[11] provides additional authority for federal support to state and local governments to both prepare for and respond to major disasters, including wildland fires. The Stafford Act establishes a process for requesting and obtaining a presidential disaster declaration, defines the type and scope of available federal assistance, and sets conditions for obtaining that assistance. The act requires that the Governor of the affected state request a presidential declaration based upon a finding that effective response is beyond the capabilities of the state and affected local governments and that federal assistance is necessary. The act authorizes the President to direct any federal agency to provide assistance—including grants, equipment, supplies, and personnel—to any state or local government for the mitigation, management, and control of any fire on public or private forest land or grassland if it threatens to become a major disaster.

## Agreements Governing the Use of Military Assets under Federal Control

In addition to the laws providing broad authority for using military aid in wildland firefighting, two agreements govern the use of military assets controlled by the federal government. The first is an agreement among DOD, the Department of Agriculture, and the Department of the Interior that outlines general guidelines, responsibilities, and reimbursement for wildland firefighting. Under this agreement DOD, consistent with defense priorities, provides assistance in the following two situations:

---

[10]Office of Management and Budget, *A Review of Existing Authorities and Procedures for Using Military Assets in Fighting Wildfires*, (Washington, D.C.: May 17, 2004).

[11]42 U.S.C. §§ 5121–5206.

- DOD can provide assistance when NIFC[12] has requested it and DOD has determined that military assistance is required and justified to suppress a wildland fire. Assistance can be requested for fires on federal, state, or private property. Requests should state that all available or suitable civilian resources have been committed and that requested support does not compete with private enterprise.

- DOD can provide assistance when a forest or grassland fire on state or private land is declared a major disaster, or a determination for emergency assistance is made by the President, and the required military support is requested by the Federal Emergency Management Agency Regional Director,[13] under the Disaster Relief Act of 1974.

This agreement between federal firefighting agencies and DOD was first signed in 1975 and is in the process of being updated, although it had not been signed, as of February 2005. According to officials, the most significant change proposed in the 2005 update is a new interpretation of the agreement's reimbursement clause, which would require federal agencies to reimburse DOD, not only for costs exceeding normal operating expenses such as those for firefighting boots, but for all costs of using military personnel, including payroll costs.

The second agreement, between DOD and NIFC, governs the use of military helicopters for transporting passengers, cargo, or water in external buckets. This agreement outlines responsibilities, operational procedures, and related issues. The agreement emphasizes that flight safety standards will not be compromised in carrying out a firefighting mission.

## Policies and Procedures Governing the Use of Federally Controlled Military Assets

Finally, both DOD and NIFC have policies and procedures providing more specific guidance governing the use of federally controlled military assets for wildland firefighting. DOD directive 3025.15 establishes DOD policy and assigns responsibilities for providing military assistance to civil authorities. Specifically, it states that DOD approval authorities evaluate all requests by civil authorities for DOD military assistance against the following criteria:

---

[12]NIFC was formerly called the Boise Interagency Fire Center (BIFC); the agency's name was changed in 1993.

[13]The agreement refers to the Federal Disaster Assistance Administration (FDAA), now part of the Federal Emergency Management Agency, Department of Homeland Security.

- legality (compliance with laws),

- lethality (potential use of lethal force by or against DOD forces),

- risk (safety of DOD forces),

- cost (who pays and the impact on DOD's budget),

- appropriateness (whether the requested mission is in DOD's interest to conduct), and

- readiness (impact on DOD's ability to perform its primary mission; defense of the nation).

The Assistant Secretary of Defense for Homeland Defense evaluates requests for DOD military assistance on the basis of these criteria to determine whether resources are available and what the impact their use for firefighting would have on military readiness. The Joint Director of Military Support determines which assets would best meet NIFC's request, and the Secretary of Defense approves the order to deploy DOD resources to the fire. NIFC officials said that DOD has normally provided requested resources.

NIFC policies and procedures are contained in two primary guides, the *National Interagency Mobilization Guide* and the *Military Use Handbook*. Together these lay out under what circumstances military assets can be used; the process for ordering these resources; training requirements for personnel, including pilots, and military personnel managing aviation assets; limitations on the use of these assets; and other operational issues. Both guides state that before military assets can be mobilized, all civilian resources must be committed to ongoing suppression efforts.

According to NICC and DOD officials, current laws, agreements, and policies and procedures for requesting military aid for firefighting have proven adequate, and the process generally works well. NICC and DOD officials meet annually to discuss any needed changes to the process or procedures. Officials said that having a military liaison on-site at NIFC, when a fire season becomes severe, is a key factor in effective communications between NIFC and DOD. A May 2004 OMB report also found that authorities and policies for using military resources to fight wildland fires have generally worked well and continue to be appropriate.

The report stated that existing authorities are being used in a manner consistent with the available capabilities of DOD assets to fight wildland fires in the most expeditious and efficacious way to minimize the risk to public safety.

## Procedures Governing the Use of Military Assets under State Control

In contrast to the process for obtaining military resources under federal control for federal firefighting purposes, the use of National Guard units under state control is outlined in memorandums of understanding among federal agencies, state agencies, and specific National Guard units. National Guard assets under state control normally do not operate outside their state boundaries. The agreements authorizing their use vary in specificity, but National Guard assets are generally deployed only after a state's governor has declared a state of emergency. The agreements or other associated documents, such as operating plans, may include the circumstances under which the assets can be used, process for requesting the assets, and training and reimbursement requirements. For MAFFS, the Forest Service develops an annual operating plan that includes this information. Procedures or operating plans governing the use of other National Guard assets, such as helicopters, are prepared by the state. In California, the Department of Forestry and Fire Protection worked with the California National Guard, the Forest Service, and the National Park Service to develop detailed operating plans and training guides for the use of military helicopters. Not all states that use these resources for wildland firefighting have developed such guidance, however.

## Military Assets Used for Wildland Firefighting 1988-2003

NIFC maintains information on the use of military assets under federal control, including military personnel, as well as MAFFS air tankers, operated by either the Air Force Reserve or National Guard. According to a NICC official, military personnel and equipment were rarely used for firefighting before 1988. From 1988 through 2003, however, severe fire seasons have resulted in the use of federal military resources or MAFFS in 9 of 16 years (see table 2).

**Table 2: Federal Military and MAFFS Assets Used for Wildland Firefighting 1988-2003**

| Year | Days in preparedness level 5[a] | Military assets |
|---|---|---|
| 2003 | 48 | 1 Army battalion with medical evacuation helicopter<br>8 Air National Guard and Air Force Reserve C-130 tankers (MAFFS)<br>6 Marine Corps helicopters<br>4 Navy Reserve helicopters |
| 2002 | 62 | 1 Army battalion<br>8 Air National Guard and Air Force Reserve C-130 tankers (MAFFS) |
| 2001 | 16 | 2 Army battalions<br>8 Air National Guard and Air Force Reserve C-130 tankers (MAFFS) |
| 2000 | 40 | 4 Army battalions<br>1 Marine Corps battalion |
| 1996 | 21 | 1 Army battalion<br>1 Marine Corps battalion<br>8 Air National Guard and Air Force Reserve C-130 tankers (MAFFS) |
| 1994 | 46 | 5 Army battalions<br>2 Marine Corps battalions<br>8 Air National Guard and Air Force Reserve C-130 tankers (MAFFS) |
| 1990 | Not available | 4 Army battalions<br>8 Air National Guard and Air Force Reserve C-130 tankers (MAFFS) |
| 1989 | Not available | 4 Army battalions<br>19 helicopters<br>8 Air National Guard and Air Force Reserve C-130 tankers (MAFFS) |
| 1988 | Not available | 6 Army battalions<br>2 Marine Corps battalions<br>57 helicopters (including 2 with infrared scanners)<br>8 Air National Guard and Air Force Reserve C-130 tankers (MAFFS) |

Source: NIFC.

[a]During 1995, 1997, 1998, and 1999, the nation never reached preparedness level 5 and no active military or MAFFS assets were used for firefighting. Information on days in preparedness level 5 was not available for 1993 or earlier.

Complete information on National Guard assets assisting federal or state wildland firefighting efforts was not readily available on a national level. National Guard helicopters, military personnel, or other resources, however, have been used in a number of states in recent years including California, Florida, Montana, and Oregon. According to a California National Guard official, National Guard helicopters have been used in each of the last 15 years to assist in wildland firefighting. The Florida Division of Forestry Air Tactical Coordinator said that Florida used National Guard helicopters and military personnel each year from 1998 through 2002.[14] Oregon has also used National Guard resources, such as during the severe 2002 fire season.

---

[14]The number of National Guard personnel used by the Florida Division of Forestry ranged from 30 to 150.

# List of Participants in the Symposium Convened for GAO by the National Academy of Sciences

Hank Blackwell
Fire Marshal
Assistant Chief
Santa Fe County Fire Department

Thomas Chirhart
SAFECOM Spectrum Program Manager
Office of Science & Technology
Department of Homeland Security

Jack D. Cohen
Research Physical Scientist, Fire Sciences Laboratory
USDA Forest Service

Ed Dickerhoof
Economist
Resource Valuation and Use Research Staff
Research and Development Division
USDA Forest Service

Doug Dierdorf
Senior Scientist, Fire Research Group
Air Force Research Laboratory

David D. Evans
Fire Protection Engineer
Building and Fire Research Laboratory
National Institute of Standards and Technology

Nicholas Flores
Associate Professor of Economics
Research Associate, Institute of Behavioral Sciences
University of Colorado

Jeffrey W. Gilman
Research Chemist
Materials and Products Group
Fire Research Division
National Institute of Standards and Technology

Jeffrey T. Inks
Assistant Staff Vice President

Appendix II
List of Participants in the Symposium
Convened for GAO by the National Academy
of Sciences

Codes and Standards Advocacy Group
National Association of Home Builders

Rich Just
Director, Fire Operations
Thermo Technologies, LLC

Paul Kleindorfer
Professor
The Wharton School
University of Pennsylvania

Judith Leraas Cook
Project Manager, Firewise Communities/USA

Chris Lewis
Office of the Chief Information Officer
Telecommunications System Division
Department of the Interior

Tara McGee
Associate Professor
Department of Earth & Atmospheric Sciences
University of Alberta

Julio "Rick" Murphy
Telecommunications Specialist
Wireless Management Office
Department of Homeland Security

Robert D. Neamy
Deputy Chief
Los Angeles City Fire Department (Retired)

Don Oaks
Viking Research
Co-Chair, California Fire Chiefs Association Urban-Wildland Committee,
Fire Prevention Officers Section Southern Division

William M. Raichle
Assistant Vice President
Insurance Services Office

**Appendix II**
**List of Participants in the Symposium**
**Convened for GAO by the National Academy**
**of Sciences**

Ronald G. Rehm
Fellow
Building and Fire Research Laboratory
National Institute of Standards and Technology

Jim Ridgell
Vice President and General Manager, Federal Business
EF Johnson

James C. Smalley
Manager, Wildland Fire Protection
National Fire Protection Association

Joe Stutler
Forestry Specialist
Deschutes County (Oregon)

Jim Tidwell
National Director, Fire Service Activities
International Code Council

Robert H. White
Project Leader, Fire Safety Research Work Unit
Forest Products Laboratory
USDA Forest Service

Joseph Zicherman
President
Fire Safety Consultant
Fire Cause Analysis

# Technologies to Protect Structures from Wildland Fires

## Fire-resistant roof-covering materials

| | |
|---|---|
| **What they do and how they are used** | A variety of noncombustible or fire-resistant materials are available to construct roofs. During a wildland fire, they protect against firebrands landing on a roof and igniting it. Noncombustible materials will not catch fire. Fire-resistant ones will not catch fire immediately but may eventually ignite. The overall fire resistance of a roof is determined by the design and construction of the entire roofing assembly, including any intermediate layers, called "underlayments," the roof decking, and the outermost layer. Roofing assemblies are evaluated according to standardized methods as class A, class B, or class C. Class A roofs are recommended for protection of structures in areas of extreme wildland fire risk, while class C roofs are recommended for areas of low risk. These fire-resistant roofing materials can be used for roofs on new homes or when roofs are replaced on existing homes. |
| **Types of roof-covering materials** | **Asphalt composition:** Fiberglass or paper mats combined with asphalt and coated with small amounts of minerals or stone. Typically available in class A or C. The most widely used roofing material and one of the most inexpensive fire-resistant roofing materials. |
| | **Clay:** Fine-grained earthy material that hardens when heated and is widely used to make bricks and tiles. Noncombustible, class A. Is more expensive than many other materials and may be too heavy for some uses. |
| | **Concrete:** Usually a mix of cement, sand, gravel, and water that can be made to look like wood shingles. Noncombustible, class A. Can cost and weigh less than clay. |
| | **Fiber-cement:** Cement combined with wood fiber that can be molded to look like shingles and shakes. Noncombustible, requires underlayment to achieve class A. May be susceptible to water damage. |
| | **Metal:** Generally steel or aluminum, available in flat sheets with seams or a finish that looks like wood. Noncombustible but requires gypsum underlayment under the outer covering to restrict heat transfer to achieve a class A rating. Lightweight and durable. |
| | **Slate:** A fine-grained rock that can be split into thin, smooth layers. Noncombustible, class A. Highly durable but more expensive than many other coverings. May require additional roof support because of its weight. |
| | **Synthetic rubber:** Often made from recycled rubber and molded to look like traditional wood or slate. Available in class A, B, or C but may need additional underlayments to achieve a specific rating. Is cheaper and can weigh less than real slate. |
| | **Treated wood:** Wood may be pressure treated with chemicals to make it fire resistant. Combustible, available in class A, B, or C but may need additional underlayments to achieve a specific rating. Fire-resistant treatment may deteriorate over time. |
| **Effectiveness** | The use of noncombustible or fire-resistant roofing materials has been shown to be a critical factor in protecting structures from wildland fire. Class A roofs are more fire resistant than class B or C roofs, but all offer some protection from wildland fire. While none will readily allow fire to spread across the roof, a noncombustible material may offer better protection. Some combustible materials depend on chemical treatments for their fire performance, and experts are concerned about whether such treatments will last the lifetime of the roof. Moreover, it is important to evaluate the entire roof assembly, not just the roof covering, when determining effectiveness. Metal, for instance, is noncombustible but can transfer heat to the materials underlying it and ignite them. |

(Continued From Previous Page)

**Fire-resistant roof-covering materials**

**Key factors affecting cost**

A number of factors can affect the cost of roof-covering materials. Asphalt composition and metal roof-covering materials are less expensive or comparably priced to untreated wood. Other roofing materials, such as concrete or clay tiles, may be more expensive and some, such as slate, may be substantially more expensive (see fig. 19). However, these costs can vary depending on the geographic location of the home.

**Figure 19: Comparison of Estimated Cost of Common Fire-Resistant Roof-Covering Materials**

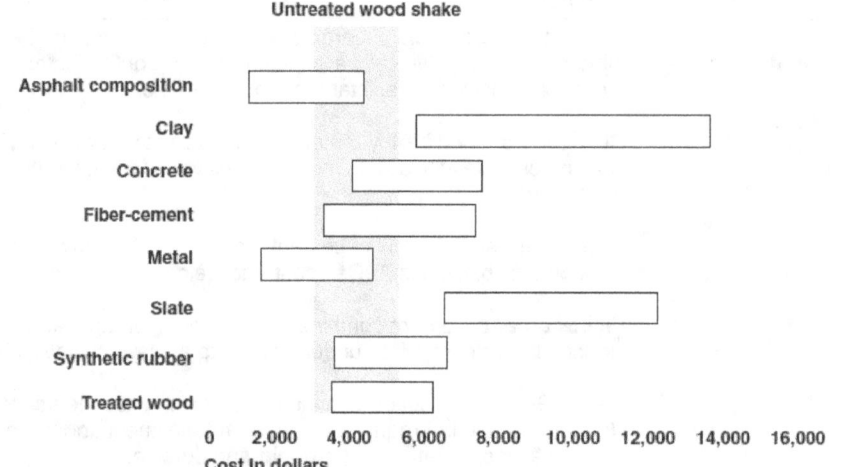

Source: GAO analysis of Marshall & Swift 2004 Residential Cost Handbook data.

Note: Using a nationally-recognized construction cost guide, we estimated the cost of roof-covering materials needed for a 2,000-square-foot, two-story home with no garage. Costs illustrated represent the cost of the roof-covering material and installation, as compared with the cost of an untreated wood shake roof. Due to the weight of some roof-covering materials, such as clay or slate, additional costs may be required to strengthen the roof structure.

## Fire-resistant windows

| | |
|---|---|
| **What they do and how they are used** | Exposure to intense heat from a wildland fire can crack a glass window, even without direct contact, and allow fire to enter a structure. Conventional glass windows may crack after approximately 5 minutes of heat exposure. A variety of fire-resistant windows are available to help protect a structure from igniting by providing more resistance to cracking. |
| **Types of fire-resistant windows** | **Dual-paned glass:** Contains two layers of glass. The first layer partially protects the second layer and roughly doubles the amount of time before a window cracks when exposed to the heat from a wildland fire. Frequently used because it increases energy efficiency.<br><br>**Glass blocks:** Most fire-resistant glass material available. Use may be limited because it allows light to enter a structure but does not provide a clear view through the glass.<br><br>**Tempered glass:** Has been strengthened to resist breaking from heat. Can also offer protection from flying debris. |
| **Effectiveness** | Fire-resistant glass provides more protection than conventional glass from the heat generated by a wildland fire. If a window does crack from exposure to heat, a smaller window is more likely to stay in place and continue to protect the inside of a structure. The frame holding a glass window in place also needs to be able to withstand fire. Aluminum frames offer more protection than wood frames, which are highly combustible, or vinyl frames, which can melt and allow the glass to fall away. Finally, metal shutters or screens can offer additional protection for windows by decreasing the duration of a window's exposure to heat. |
| **Key factors affecting cost** | A variety of factors affect the cost of windows, including glass type, style, size, quality, and framing materials. |

## Fire-resistant building materials

| | |
|---|---|
| **What they do and how they are used** | A variety of noncombustible and fire-resistant materials are available to construct exterior components, such as walls, siding, and doors. These materials protect against flames or intense heat igniting or melting away a structure's exterior. They can also be used to construct such things as decks and fences which, if ignited, can lead fire to the dwelling. Noncombustible building materials will not catch fire, and fire-resistant ones will not catch fire for a period of time but may eventually ignite. The overall fire resistance of a building component is often determined by the length of time its entire assembly can contain a fire or maintain its structural integrity against fire. This fire performance is often rated according to standardized methods as 20-minute, 1-hour, 2-hour, or 4-hour. |
| **Types of fire-resistant building materials** | **Fiber-cement:** Cement combined with wood fiber. Available in a wood-grain finish. Noncombustible but may need an underlying gypsum sheathing to achieve a 1-hour rating. |
| | **Heavy timber:** Combustible, but the low surface-to-volume ratio of thick timbers—typically, a minimum thickness of 6 inches for exterior siding—causes them to resist ignition and burn slowly. Very durable. |
| | **Masonry:** Brick, stone, or block. Noncombustible, usually 2-hour rated. Very durable. |
| | **Metal:** Metal siding—generally steel or aluminum—available in flat sheets with seams or a finish that looks like wood. Noncombustible but requires an underlying gypsum sheathing to achieve a 1-hour rating. |
| | **Plastics and wood-plastic composites:** Plastics, sometimes combined with natural wood fiber, that can be manufactured to look like wood. Used mainly for decking and fences. Low combustibility. |
| | **Stucco:** Plaster typically made of cement, sand, and lime, applied in two or three coats over a metal reinforcing mesh to form a three-fourths-inch to one-inch finished layer. Stucco can be colored and scored to appear like brick, stone, or other materials. Noncombustible, 1-hour rated. It can be prone to cracking if not applied correctly. |
| | **Treated wood:** Wood may be pressure treated with chemicals to make it fire-resistant. Combustible. Fire-resistant treatment may deteriorate over time. |
| **Effectiveness** | The use of noncombustible or fire-resistant building materials has been shown to be helpful in reducing a structure's vulnerability to wildland fire. Longer-rated materials offer more protection than shorter-rated materials. Time-dependent ratings, however, may not be meaningful in wildland fires because firefighters may not be able to respond for many hours, if at all. Noncombustible materials can offer better protection. As with roofs, it is important to evaluate the entire assembly, not just the outer layer of material, when determining effectiveness. Some materials require additional layers to achieve a particular fire performance. |
| **Key factors affecting cost** | Estimated costs of fire-resistant building materials vary widely. Using a nationally-recognized construction cost guide, we estimated the construction cost of a 2,000-square-foot, two-story home with no garage or basement. The estimated construction costs for a wood-framed home using wood, metal, or stucco exterior building materials were comparable. The estimated costs using brick exterior building materials was about 10 percent more, and stone was about 20 percent more. For decking material, the cost of plastic and composite materials is comparable to the higher-end wood products, such as redwood, but more expensive than treated wood. However, these costs can vary depending on the geographic location of the home. |

| Chemical agents | |
| --- | --- |
| **What they do and how they are used** | Chemical agents are used with water to provide a temporary protective coating that inhibit ignition of flammable surfaces. They are designed to overcome some of water's drawbacks, including its tendency to bead and to run off vertical surfaces. Chemical agents can be applied by firefighters or by homeowners. Homeowners can apply them using plastic containers attached to a standard garden hose or using portable pump systems. Permanently installed units are also available. These systems are often provided with their own water and power, and some can be set up to distribute the agent to nozzles mounted on the roof. |
| **Types of chemical agents** | **Foams**: A mass of air-filled bubbles formed by forcibly mixing water and a wetting agent with air. Often composed of ingredients found in natural or synthetic detergents, such as dishwashing liquid or shampoo. |
| | **Gels**: Superabsorbent molecules (polymers) that retain hundreds of times their weight in water. They adhere well to vertical surfaces such as walls. |
| | **Wetting agents**: Surfactants (surface active agents) that reduce water's surface tension, increasing its ability to permeate a surface. Often a component of a foam or gel. |
| **Effectiveness** | Chemical agents have been shown to be effective in temporarily protecting structures from fires. These agents increase the efficiency of water as a firefighting tool, reducing the amount of water needed for effective suppression. For example, research at the University of Toronto has shown that coating structures with surfactants can reduce the amount of water needed to fight a fire by as much as 60 percent. Unlike passive protection systems such as fire-resistant building materials, application of chemical agents typically requires either firefighters or homeowners to be present. In addition, foams and gels may dry out before the wildland fire risk has passed and need to be reapplied. They are not effective once the water has evaporated. Further, once applied, gels can be difficult to clean up and may require multiple washings to remove after a fire has passed. The Forest Service maintains a qualified list of wildland fire chemical agents that have been tested against environmental and health standards. |
| **Key factors affecting cost** | A variety of factors affect the cost of chemical agents, including whether the system used to apply chemical agents is portable or installed and whether power and water are supplied with the system. |
| | The cost of systems to apply chemical agents varies widely depending on features. These systems can cost more when power and water are supplied with the dispensing systems. |

| Sprinkler systems | |
| --- | --- |
| **What they do and how they are used** | Sprinkler systems spray water on the inside or outside of a structure. Some external sprinklers can also spray chemical agents. |
| **Types of sprinkler systems** | **Interior sprinklers:** Often used to protect from more-typical structural fires—such as those caused by cooking, smoking, or other hazards—but also offer protection from fires that start with firebrands entering a house through a vent or other opening, especially if the sprinkler is mounted in the attic. Frequently activated automatically.<br><br>**Exterior portable sprinklers:** Some can be attached directly to a garden hose or to a small portable pump to increase water pressure. Some can be placed on the roof.<br><br>**Exterior permanent sprinklers:** Permanently installed systems that often require large sources of water. One such system includes retractable roof-mounted sprinkler nozzles that emerge when needed and retract when not in use. Some can be activated automatically. |
| **Effectiveness** | Sprinkler systems provide additional protection for structures by decreasing a structure's flammability and reducing the chance of ignition. Exterior sprinklers can also decrease the flammability of nearby vegetation, further increasing protection. Sprinkler systems, however, may be ineffective in a wildland fire because of shortages of water or power. In addition, temporary sprinkler systems require homeowners to be present to set up and activate them. |
| **Key factors affecting cost** | The cost of sprinkler systems varies considerably depending on whether the system is interior or exterior, permanent or portable. Advanced features, such as automated detection and activation, can also affect the cost. |

## Geographic information systems

| | |
|---|---|
| **What they do and how they are used** | Geographic information systems (GIS) are a computer-based information system for storing, analyzing, and displaying complex information. GIS links sets of data and displays the information as maps with many different layers, each representing a particular "theme," or feature. For example, one theme could map all the homes in a specified community, another could map the streets in the same area, and still others could map vegetation or water resources. Analyzing the relationships among these features can significantly aid decision makers with complex choices, such as where to place new fuel breaks. |
| **Effectiveness** | GIS has been shown to be an effective tool for community planning to protect structures and communities from wildland fires. GIS allows fire officials to analyze vegetation distribution, predicted fire behavior, and location of structures to identify areas most at risk. This information can be used to determine where action— such as vegetation management, fuel breaks, or educational outreach programs—is most needed. For example, the Los Angeles County Fire Department uses GIS to identify high-risk areas within its jurisdiction and then assesses its resources and prescribes vegetation management accordingly. |
| **Key factors affecting cost** | The cost of GIS systems varies widely, depending on the system and scope of use. The cost associated with collecting and maintaining data for GIS use can be substantial. Some GIS systems offer public access to data on the Internet without charging access fees to users. |

Source: GAO analysis of federal, state, local, nongovernmental, and commercial data.

# Web Sites with Information on Protecting Homes from Wildland Fire

Federal Emergency
Management Agency
www.fema.gov/hazards/fires/wildfires.shtm

Firewise Communities
www.firewise.org/

Forest Service Database of
Wildland Fire Mitigation Programs
www.wildfireprograms.com/

Florida Demonstration Home
www.interfacesouth.org/fire/firewisehome/

National Association of State Foresters
www.stateforesters.org/

National Fire Plan
www.fireplan.gov/

National Interagency Fire Center
www.nifc.gov/

# Technologies for Improving Communications Interoperability

Firefighting and other public safety personnel responding to wildland fires need to be able to communicate with one another. The ability of any public safety official to talk to whomever they need to, whenever they need to, is commonly called communications interoperability. Many agencies, however, either operate on different radio frequency bands or use incompatible communications systems. Technologies are currently available, and others are being developed, to help public safety agencies overcome these barriers. These technologies can be grouped into short-term, or patchwork, solutions to interconnect existing radio systems and longer-term solutions to upgrade communications systems for increased interoperability.

## Patchwork Interoperability

Patchwork interoperability uses technology to interconnect two or more disparate radio systems so that voice or data from one system can be made available to all systems. A key advantage of this solution is that it can tie together existing communications systems and requires only minimal additional equipment. Three primary patchwork solutions exist.

## Audio Switch

An audio switch provides interoperability by sending audio from one radio system to all other connected systems. An audio switch can be either stationary or mobile. One popular audio switch (see fig. 20) consists of a chassis with slots, into which different hardware modules can be installed to control and interconnect different communications systems, such as VHF (very high frequency) and UHF (ultrahigh frequency) radios, as well as telephones. The audio switch can hold up to 12 interface modules, each capable of connecting a radio system. Further, two chassis can be linked, doubling the number of radio systems that can be connected.

**Figure 20: An Audio Switch**

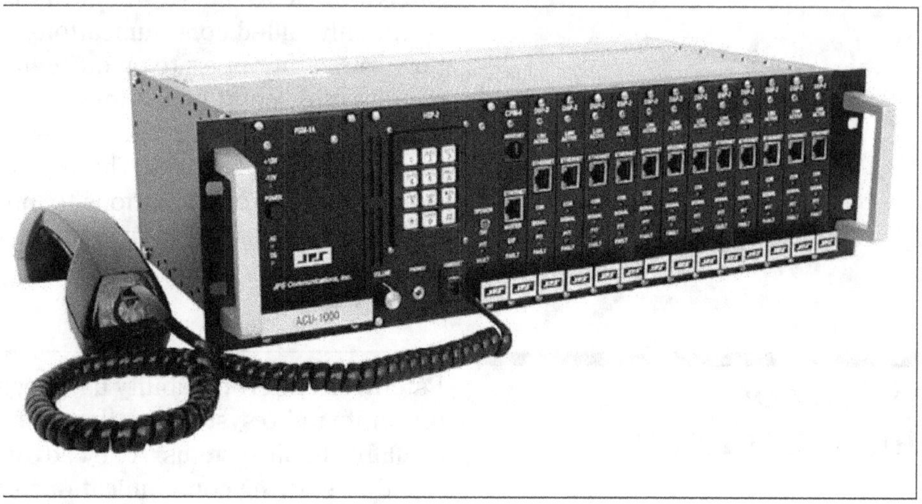

Source: Raytheon JPS Communications.

Audio switches are useful for wildland firefighting where multiple agencies temporarily come together to fight the fire because they are easily transportable and can be used to create temporary interoperability. A portable audio switch is available for easy transport. Audio switches also provide flexibility because different agencies can be connected in different incidents or situations, although a different type of cable is needed for each type of radio connected. Finally, audio switches may cost less than some other interconnection devices, such as crossband repeaters, although audio switches still may be out of reach of agencies facing funding constraints. For example, one audio switch costs approximately $7,000 for the chassis without the radio interface modules or cables. An interface module and a cable are needed for each radio connected. The module costs approximately $1,100, and the cables are available for approximately $140 each.

Audio switches are relatively new. According to an official with the National Interagency Fire Center's (NIFC) National Interagency Incident Communications Division, which maintains the nation's radio cache, has acquired two audio switch units that will be available to firefighting agencies for the first time in the 2005 fire season.

## Crossband Repeater

A crossband repeater provides interoperability between systems operating on different radio frequency bands by changing frequencies between two radio systems. Crossband repeaters can connect base stations[1] or mobile radios, either in vehicles or handheld (see fig. 21). The repeater is also useful for extending the communications coverage beyond the range of a single radio. Crossband repeaters can also be linked together to overcome distances or geographical features blocking communication among users utilizing one repeater.

**Figure 21: A Crossband Repeater Used to Connect Radios Operating on Different Frequency Bands**

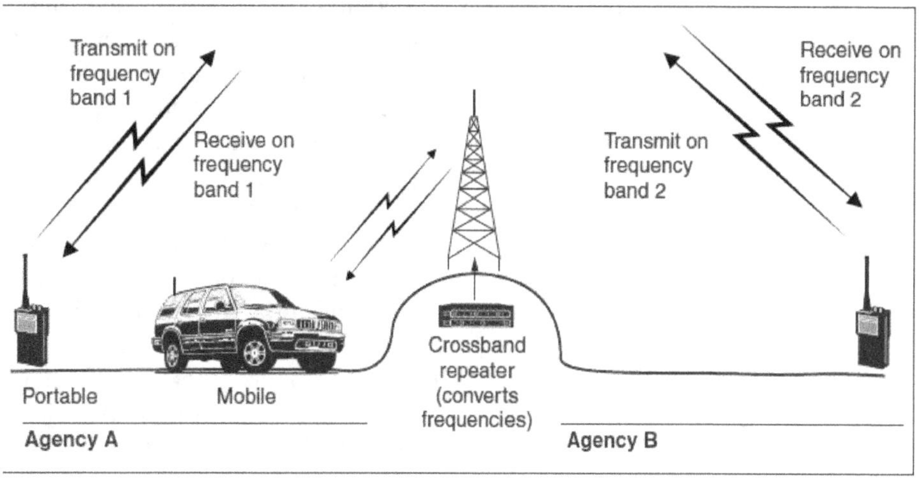

Sources: GAO, DHS, and Nova Development Corp.

According to a communications specialist at NIFC, crossband repeaters are an effective interoperability solution often used by federal firefighting agencies. For example, federal firefighting agencies operate on both VHF and UHF when fighting a wildland fire. VHF (AM and FM) is used for tactical communications by personnel at the fire line and tactical aircraft flying over the fire and UHF (AM) is used in the base camp for logistical or other nontactical uses. When federal firefighting agencies are at an incident, a crossband repeater can be temporarily set up on a nearby hilltop

---

[1]A base station contains the equipment for transmitting and receiving the radio signals that allow portable radios to communicate with each other.

to transmit signals between these different frequency bands. The device receives a VHF signal and retransmits it as a UHF signal, for example. NIFC has crossband repeaters available and can quickly transport them to the incident. Ranging in price from $4,000 to $33,000 each, crossband repeaters can cost more than audio switches, which may put them beyond the reach of small state and local jurisdictions with limited funding.

## Console-to-Console Patch

A console-to-console patch achieves interoperability by making an audio connection between the dispatch consoles of two different radio systems. Unlike patchwork solutions that can be transported to the emergency or incident, console-to-console patches connect consoles located at the dispatch centers where personnel receive incoming calls. These patches can connect personnel from an agency using one radio system to personnel from an agency using a different radio system. Connections between dispatch consoles can be made temporarily, as needed, through a public telephone line or permanently over a dedicated leased line or a dedicated microwave or fiber link.[2] The costs for this type of solution primarily depend on the type of link used. For example, the costs for a console-to-console patch over a telephone line or a dedicated leased line are fairly minimal and would primarily consist of the recurring telephone line charges. In contrast, dedicated microwave or fiber links require a significant initial investment. For example, a typical microwave link connecting two locations about 3 to 5 miles apart would require an initial investment of around $35,000 whereas connecting two locations about 15 to 25 miles apart would double the investment to about $70,000. Figure 22 illustrates the concept of a console-to-console patch over a dedicated link.

---

[2]A leased line refers to a permanent telephone connection set up by a telecommunications provider between two geographic locations. A fiber link uses light sent over a glass or plastic fiber to carry communication signals. A microwave link uses radio beams of extremely high frequencies to send information between two fixed geographic sites.

Figure 22: Console-to-Console Patch over a Dedicated Link

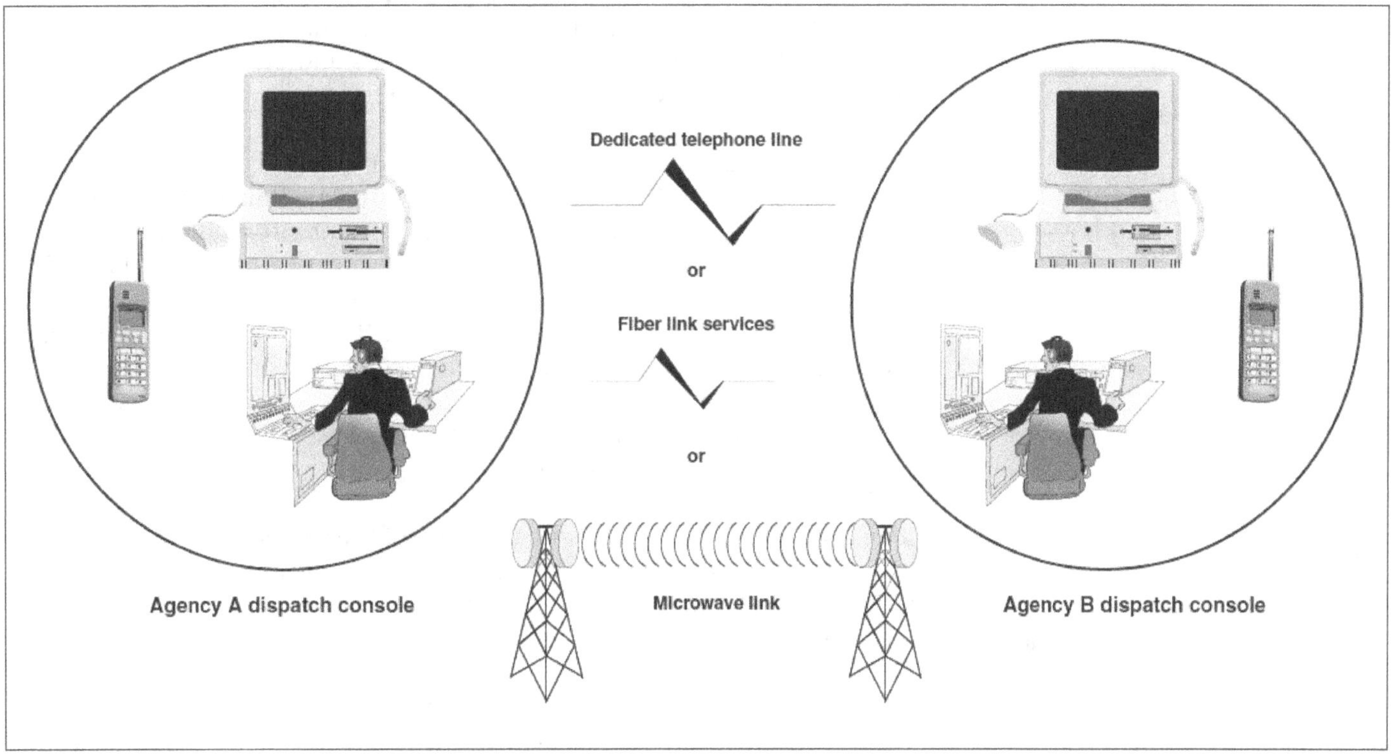

Sources: GAO, DHS, and Nova Development Corp.

# Improved Communication Systems

Beyond patchwork solutions, improved interoperability can also be achieved by adopting better communications systems that use a set of common technical standards or provide more sophisticated communications capabilities. These new technologies require replacing or gradually phasing out existing radio systems.

# Project 25 Systems

Project 25, also called APCO 25, was begun in 1989 by representatives from the Association of Public Safety Communications Officials International, the National Association of State Telecommunications Directors, the National Telecommunications and Information Administration, the National Communications System, and the Department of Defense, to provide detailed standards for digital two-way wireless communications systems so that all purchasers of Project 25-compatible equipment can communicate with each other. Project 25 has two main phases. During the first phase, five standards were completed and published. Equipment compatible with these standards are available from multiple vendors. Phase 2 of the project focuses on developing standards for other components of the systems, such as dispatch consoles and base stations.

Project 25 radios provide several benefits for users. First, they can carry both voice and data. This feature can be useful in wildland firefighting because it can provide firefighters with important information about subjects such as weather or fire behavior. Second, Project 25 digital radios can operate in narrowband frequencies, which allow more users within the existing public safety radio frequency bands. Current analog public safety radios use 25 kHz-wide channels for each conversation. Project 25 radios use 12.5 kHz-wide channels, so that two conversations can take place in the space where only one used to fit. Eventually, these radios will use 6.25 kHz-wide channels, allowing four times as many conversations as analog radios. At the same time, however, Project 25 radios are "backward compatible" so they can still communicate with analog radios and operate in analog mode on 25 kHz channels. This backward compatibility enables agencies to make the transition to digital Project 25 radios gradually, while continuing to use their analog equipment.

While Project 25 radios provide additional capabilities, they are also more costly, which is a barrier for many public safety agencies with limited funding. For example, Project 25 portable radios, priced between $1,700 and $2,500, cost more than other available radios that cost around $1,200 each.

Although the federal government has begun moving to Project 25 standards, it will take several years for the federal government to replace all existing radios with Project 25 radios. According to federal officials, the Department of the Interior and the Forest Service did not adopt Project 25 radio standards at the same time. In 1996, the Department of the Interior adopted both narrowband and Project 25 digital standards. According to an official, by October 2004, the department had converted just over half of the communications infrastructure to Project 25 digital technology. In contrast, the Forest Service initially adopted narrowband analog, but not digital, standards. The Forest Service completed the migration to narrowband by January 2005, according to a Forest Service official and, in October 2004, began requiring that all new radios purchased meet Project 25 digital standards. Full implementation of Project 25 within the Forest

Service is not expected to take place until about 2012. In the case of NIFC, it has 1,500 Project 25 radios in its cache, out of a total of 8,000 radios, according to a NIFC communications specialist.

It will also take time for other jurisdictions to migrate to these radios. For example, a December 2003 inventory of communications equipment in Washington State showed that about 400 state-owned portable radios are Project 25 compatible, however, none of these are owned by the Department of Natural Resources, which is responsible for wildland firefighting.

## Software-Defined Radios

Software-defined radios, first developed by the Department of Defense (DOD),[3] are an emerging technology that holds potential for public safety agencies, including firefighting agencies. These radios use software to determine operating parameters such as the frequency band (such as VHF or UHF) and modulation type (such as AM or FM). Because these parameters are determined by software, a software-defined radio could be programmed to transmit and receive on any frequency and to use any desired modulation within the limits of its hardware design. Software-defined radios will allow interoperability between agencies using different frequency bands, different operational modes (digital or analog), proprietary systems from different manufacturers, or different modulation (AM or FM). For example, a software-defined radio can be programmed to work as a conventional UHF radio but in another operating mode can function as an 800 MHz radio. Some software-defined radios could be used to identify unused frequencies and automatically make use of them, which is important in making efficient use of limited radio spectrum. The software-defined radio technology may also provide integrated voice and data over the same channel, a useful feature for firefighters who need maps, weather, and fire behavior data. These radios, however, are still being developed and are not yet available for use by public safety agencies.

## Voice over Internet Protocol

Voice over Internet Protocol can connect different radio systems by using the Internet as the connecting mechanism. Voice over Internet Protocol converts analog voice signals from a radio into digital data packets that

---

[3]This technology began within DOD's SPEAKeasy research program in 1992 and beginning in 1997 became part of DOD's Joint Tactical Radio System Program.

travel over an Internet Protocol network.[4] At their destination, the digital information is converted back to analog audio and can be heard on the recipient's radio. Voice over Internet Protocol enables interoperability between agencies using different frequency bands, different operational modes (digital or analog), or proprietary systems from different manufacturers.

Voice over Internet Protocol holds promise as a relatively low-cost solution to communications interoperability, but some key issues will need to be resolved before it can be an effective solution. It works using Internet Protocol, which is a widely used technology standard, so commercial off-the-shelf network equipment is available from many vendors which keeps costs relatively low. However, no standards exist for radio communications using Voice over Internet Protocol and, as a result, manufacturers have produced proprietary systems that may not be interoperable. In addition, the system does not yet have reliable voice quality. During periods of network congestion, packets of voice information can be distorted or dropped. A communications specialist with NIFC explained that while data packets can be sent again, normally without adverse consequences, delayed or dropped voice communication packets can mean that personnel on the receiving end of the communication may not hear all critical information and this could put firefighters' safety at risk.

---

[4]In some cases, this is the Internet; and in others, it is a private data network.

# Comments from the Department of Agriculture

| United States Department of Agriculture | Forest Service | Washington Office | 1400 Independence Avenue SW Washington, DC 20250 |
|---|---|---|---|

**File Code:** 1310/1430
**Date:** MAR 3 1 2005

Ms. Robin M. Nazzaro
Director, Natural Resources and the Environment
U.S. Government Accountability Office
441 G Street, N.W.
Washington, DC 20548

Dear Ms. Nazzaro:

Thank you for the opportunity to review and comment on the draft Government Accountability Office (GAO) report, GAO-05-380, "Technology Assessment: Protecting Structures and Improving Communications during Wildland Fires". The Forest Service concurs with the facts presented in the report and believes that it provides an accurate, balanced assessment of the challenges and various efforts underway to protect structures and improve communication during wildland fires.

We look forward to working with GAO on future reviews. If you have any technical questions regarding this review, please contact Tory Henderson, Equipment and Chemical Branch Chief, at (208) 387-5348. For general questions, please contact Sandy T. Coleman, Agency Audit Liaison, at (703) 605-4940.

Sincerely,

DALE N. BOSWORTH
Chief

cc: Christine Roye

**Caring for the Land and Serving People**

Printed on Recycled Paper

# Comments from the Department of Commerce

**UNITED STATES DEPARTMENT OF COMMERCE**
**The Deputy Assistant Secretary for**
**Technology Policy**
Washington, D.C. 20230

Postmarked April 7, 2005 N.B.

Mr. Naba Barkakati
United States Government Accountability Office
Washington, DC 20548

Dear Mr. Barkakati:

Enclosed are comments from the Department of Commerce to the Government Accountability Office (GAO) proposed report entitled Technology Assessment: Protecting Structures and Improving Communications during Wildland Fires (GAO-05-380). Thank you for giving us an opportunity to review the report before it is issued in final form. I commend the GAO for this study on an issue of such national importance, protecting structures during wildland fires.

The authors have done an excellent job of describing the problem, assessing the vulnerabilities of individual structures to wildland fires, and developing recommendations for (1) the protection of structures subjected to wildland fires, and (2) development of technologies to assist Federal agencies working together more effectively during the fires.

We recognize that developing new physics-based computer models for the wildland-urban interface would greatly assist communities in (1) assessing fire risk, (2) developing new fire mitigation/prevention strategies, (3) evaluating the cost effectiveness of risk-reduction changes, and (4) designing improved fire safety and evacuation routes.

Again, thank you for the opportunity to comment on this draft report.

Sincerely,

Daniel W. Caprio

Daniel W. Caprio

Enclosure

**Comments on**
**Government Accountability Office (GAO) Report entitled "Technology Assessment:**
**Protecting Structures from and Improving Communications during Wildland Fires"**
**made by the**
**National Institute of Standards and Technology, Department of Commerce**

1. The GAO team should be commended for the study. The report provides a very good assessment of the vulnerability of individual structures to wildland fires, and how this vulnerability can be substantially reduced by some relatively simple and inexpensive changes. It also makes a good assessment of the communication technologies needed so there can be effective action between agencies during wildland fires.

2. The authors make an important distinction between the terms "wildfires," and "wildland-urban-interface fires" or "WUI fires." Generally, these two sets of terms are used imprecisely, and therefore, in a potentially confusing way. Wildfires are simply fires involving only wildland, or vegetation fuels. In contrast, the wildland-urban-interface (WUI) is carefully defined as an area where wildland materials (trees, shrubs and grass, for example) reside intermingled with structures. WUI fires are ones that burn in a heterogeneous environment. This definition also implies that there may be multiple structures, as well as multiple trees, shrubs, etc., in the area of concern. The distinction between "wildfires" and "WUI fires" becomes very important when the discussion turns to the physics of fire spread in WUI fuels. In contrast to wildfires, where the fuel bed can be considered statistically homogeneous, in WUI fires, the fuel is heterogeneous and occurs in discrete elements such as individual structures, shrubs, and isolated trees--each of which burns in a unique manner. The energy content and heat release from a structural fire are normally much greater than for an equivalent area of forest land.

3. This report addresses a very serious issue. The philosophy of land management within the United States has undergone substantial change during the past few decades. During most of the twentieth century, fire in wildland areas was regarded as harmful and therefore, land-management practice attempted to suppress these fires completely. The "Smokey-the-Bear" campaign was designed to reduce wildfires to the greatest extent possible.

   Fire in wildlands today is regarded as natural and so-called "prescribed burns" are used as a tool to help keep these areas healthy. However, decades of accumulated fuels in wildland areas, that would have been removed naturally by fire, have left these areas vulnerable to large, very dangerous fires. Therefore, at a time when new homes are increasing the area associated with the WUI, the results of past land-management practices have left these areas at risk to extreme, potentially catastrophic fires.

4. Revolutionary advances in computer hardware at ever decreasing costs, coupled with the development of advanced computational algorithms and a better understanding of phenomena, have changed computations in science and engineering. These advances have opened the door to the development of a new generation of models for predicting

the behavior of wildland and WUI fires. However, the development of such new-generation models still would require a multi-person, multi-year research program since these models are fundamentally different from current so-called "operational models," which are the tools used regularly to predict behavior of wildfires. Unfortunately, the research and science incorporated into these operational models is based on decades-old understanding of fire phenomena.

To be of practical predictive value, numerical simulations must be a physics-based model and developed specifically for outdoor fires, including considerations of vegetation, structures, topography and meteorology. Computer models must be able to describe fire from first principles, including use of current Geographic Information Systems (GIS) technology, and promises to be much better able to predict fire behavior under realistic outdoor conditions. The research needed for such a new model must also include a substantial experimental program to generate new data for model validation.

Over the past few decades, NIST has carried out an active experimental research and model development program related to fires *within* buildings. This research has produced new generation models that are highly regarded and useful for the prediction of building fires, outdoor smoke plumes, and other fire phenomena. The cost to develop one model, the Fire Dynamics Simulator (FDS), has been estimated to be between 100 and 200 person-years and can be used as a guide to estimate the level of research required to develop such a new generation model for outdoor fires. The National Institute of Standards and Technology simulation models have been freely distributed (http://www.fire.nist.gov/) and have been used for hundreds of applications. A recent and very important application of FDS has been its use to determine the contribution of fire to the collapse of the World Trade Center Towers.

Physics-based models can play an important role in:
    -assessing community fire risk
    -developing new fire mitigation/prevention strategies
    -testing existing assessment methodologies
    -community outreach and education
    -evaluating the cost effectiveness of risk-reduction changes
    -community design to improve fire safety and evacuation routes, and
    -predicting the smoke-plume trajectory from an approaching wild fire,
so that responders can choose the best evacuation routes.

See comment 1.

5. Key conclusions of this report are that "the two most effective measures for protecting structures from wildland fires are: (i) creating and maintaining a buffer around a structure – called a defensible space – by eliminating or reducing trees, shrubs, and other flammable objects within an area of from 30 to 100 feet around the structure, and (ii) using fire-resistant roofs and vents." The recommended defensible-space distance is based on post-fire observations and a very limited set of large-scale experiments, covering only a very restricted range of conditions related to winds, number of structures, topography, etc., that are present in WUI fires. Additional full-scale, single

See comment 2.

structure-burn experiments should be conducted but will be permitted rarely and always will be expensive. Full-scale, multiple-structure experimental burns probably will never be permitted. A new-generation physics-based simulation model, such as the one described above, could augment the limited set of large-scale experiments and serve as a cost-effective means of developing more robust recommendations for protection of structures in the WUI.

6. The communications section of the report addresses the problems related to hardware interoperability, but does not address issues of information interoperability. Examples of questions related to information interoperability are: What kinds of information do fire fighters require to deal effectively with WUI fires? What type of information should be incorporated into GIS displays and how should it be presented? What are the real-time information requirements for managing and mitigating a WUI fire? What information standards are required for seamless information flow to multiple responding organizations?

The following are GAO's comments on the letter from the Department of
Commerce, postmarked April 7, 2005.

## GAO Comments

1. Federal, state, and local materials designed to educate homeowners
   and local officials, including those published by the Firewise
   Communities program, and researchers and fire officials we spoke
   with, indicated that 30 to 100 feet of defensible space is generally
   sufficient to protect structures from wildland fire. In determining the
   amount of defensible space needed in a particular location, it is
   important to consider factors such as terrain, type of vegetation, and
   the structure's construction. Fire officials told us that, in many cases,
   local fire officials can assist homeowners in determining the
   appropriate amount of defensible space needed in their particular
   location.

2. A discussion of information interoperability for wildland firefighting
   was outside the scope of our study.

# Comments from the Department of Defense

**DIRECTOR OF DEFENSE RESEARCH AND ENGINEERING**
3030 DEFENSE PENTAGON
WASHINGTON, D.C. 20301-3030

Ms. Robin Nazzaro
Director, Natural Resources and Environment
U.S. Government Accountability Office
Washington, D.C. 20548

APR 1 3 2005

Dear Ms. Nazzaro:

This is the Department of Defense (DoD) response to the GAO draft report, "TECHNOLOGY ASSESSMENT: Protecting Structures and Improving Communications during Wildland Fires," dated March 15, 2005 (GAO Code 360474/GAO-05-380).

The draft report has been reviewed for technical accuracy and is deemed sufficient in its description of processes, usage and types of military assets to fight wildland fires. The Department concurs with the report's observation that the current laws, agreements, policies and procedures for requesting military aid for firefighting have worked well and remain appropriate.

Sincerely,

*Charles J. Holland*
For

Ronald M. Sega

**DIRECTOR OF DEFENSE RESEARCH AND ENGINEERING**
3030 DEFENSE PENTAGON
WASHINGTON, D.C. 20301-3030

APR 1 3 2005

Mr. Keith Rhodes
Director, Center for Technology and Engineering
U.S. Government Accountability Office
Washington, D.C. 20548

Dear Mr. Rhodes:

    This is the Department of Defense (DoD) response to the GAO draft report, "TECHNOLOGY ASSESSMENT: Protecting Structures and Improving Communications during Wildland Fires," dated March 15, 2005 (GAO Code 360474/GAO-05-380).

    The draft report has been reviewed for technical accuracy and is deemed sufficient in its description of processes, usage and types of military assets to fight wildland fires. The Department concurs with the report's observation that the current laws, agreements, policies and procedures for requesting military aid for firefighting have worked well and remain appropriate.

              Sincerely,

           for Charles J. Holland

           Ronald M. Sega

# Comments from the Department of Homeland Security

Note: GAO comments
supplementing those in
the report text appear
at the end of this
appendix.

U.S. Department of Homeland Security
Washington, DC 20528

Homeland
Security

March 31, 2005

Mr. Barry Hill
Director, Natural Resources and Environment
U.S. Government Accountability Office
Washington, DC 20548

Dear Mr. Hill:

     Re: Draft Report GAO-05-380, Technology Assessment Protecting Structures
and Improving Communications during Wildland Fires.

Thank you for the opportunity to review and comment on your draft report. The report is
factual and accurate providing a good assessment of the challenges facing the wildland
fire community in their efforts to protect structures and other improvements prior to a
wildfire event. The issue of improving communications during wildland fires has been a
problem for many years as frequency availability decreases with greater demand and
the need for technological solutions increases.

See comment 1.

The issue of non-compatible communications is primarily during the initial and extended
attack phases of an incident. While this is a critical period technological solutions would
best fit during this period. Large fire operations can be supported by the state, regional,
and national radio cache systems.

With the increase of homes in the wildland urban interface coupled with the severe
climate conditions due to drought, the wildland fire agencies have to divert their efforts in
perimeter control to structure protection. This ultimately leads to larger incidents and
higher cost of suppression. This report is a good focus on the actions that can be taken
prior to the incident which over time will reduce cost and fire size.

See comment 2.

In the GIS mapping discussion it should be noted that a primary use for this tool is preplanning for evacuations and public education.

As stated in the report the responsibility for land use planning resides with states and
local government authorities. While this is true it should be noted that if this is not done
well the effect is that the Federal Wildland Fire Agencies become involved in the
suppression effort both from assisting other local fire agencies as well as protecting
federal lands from fires starting on local jurisdictions.

www.dhs.gov

The Department's Science and Technology Directorate's Office for Interoperability and Compatibility (OIC) continues to refine and enhance the SAFECOM Program.

Since the release of v1.0 of the Public Safety Communications and Interoperability Statement of Requirements (SoR), SAFECOM has been in the process of developing v1.1 of the SoR. SoR v1.1 will reorganize the requirements contained within v1.0 into a layered structure, reclassifying the requirements into Network Functional Requirements, Device Functional Requirements, and Application/Services Functional Requirements. SAFECOM is currently vetting v1.1 of the SoR with the public safety practitioner community and anticipates releasing v1.1 to the public by June 30, 2005.

Development of v2.0 of the SoR is currently underway. SoR v2.0 will add additional quantitative values to the functional requirements contained in v1.1, as well as addressing National Incident Management System (NIMS) compliance. SAFECOM anticipates that it will be able to vet the draft of this version with the public safety community beginning in early 2006.

Additionally, SAFECOM awarded the contract to develop and execute the nationwide interoperability baseline study in January 2005. The purpose of the study is to quantify the extent to which the nation's public safety first responders are interoperable technically and operationally. Throughout the study, SAFECOM will utilize public safety practitioner input and analytical review by chartering a Baseline Practitioner Working Group. When it is complete, the baseline will provide understanding of the current state of interoperability nationwide. In addition, the baseline will serve as a tool to measure future improvements made through local, state, and federal public safety communications initiatives. Through the baseline, SAFECOM will be able to identify areas needing additional resources for interoperability, track the impact of federal programs and measure the success of these programs, establish an on-going process and mechanism to measure the state of interoperability on a recurring basis, and develop an interoperability baseline self-assessment tool for local and state public safety agencies. SAFECOM anticipates that it will complete the National Interoperability Baseline by December 30, 2005.

SAFECOM also recently developed the Statewide Communications Interoperability Planning (SCIP) Methodology. SAFECOM partnered with the Commonwealth of Virginia to develop a strategic plan for statewide communications and interoperability. The locally driven approach used to develop this plan can serve as a model for any state or region interested in developing a strategic plan for interoperability.

Thank you again for the opportunity to review the draft report.

Sincerely,

Steven J. Pecinovsky
Acting Director
Departmental GAO-OIG Liaison

The following are GAO's comments on the letter from the Department of Homeland Security, dated March 31, 2005.

## GAO Comments

1. We have revised the text to clarify that problems with communications interoperability occur primarily during the early stages of fire suppression efforts, called the initial and extended attack phases of the incident, before radio caches can be deployed.

2. We revised the text to clarify that GIS can also be used for community education efforts. The issue of preplanning for evacuations during wildland fires, while outside the scope of our study, was mentioned in the footnote citing previous GAO work on the uses of GIS. For more information on GIS applications for wildland fire management, see our report *Geospatial Information: Technologies Hold Promise for Wildland Fire Management, but Challenges Remain* (GAO-03-1047).

# Comments from the Department of the Interior

**United States Department of the Interior**

OFFICE OF THE ASSISTANT SECRETARY
POLICY, MANAGEMENT AND BUDGET
Washington, DC 20240

APR 1 2 2005

Robin M. Nazzaro, Director
Natural Resources and Environment
United States Government Accountability Office
441 G Street, N.W.
Washington, DC 20548

Dear Director Nazzaro:

The Department of the Interior is in substantial agreement with the major findings of the GAO draft report, *"Technology Assessment: Protecting Structures and Improving Communications During Wildland Fires,"* (GAO-05-380) (Job Code 360474). The principal findings (1. defensible space and fire-resistant roofs and vents are key to protecting structures; 2. time, expense and other competing concerns limit the use of protective measures for structures, but efforts to increase their use are under way; and, 3. effective adoption of technologies to achieve communications interoperability requires better planning and coordination) are consistent with the views of wildland fire management professionals.

The first two findings are particularly consistent with the views of Secretary Gale A. Norton, who has devoted considerable time and attention to the importance of educating the public on measures they can take to help protect their homes and property from catastrophic wildfires. Cooperation and collaboration with the public is a central focus for this Department's comprehensive approach to fire management. Community Wildfire Protection Plans need to be considered in the hazardous fuels reduction project selection process. Federal fire prevention efforts will be increasingly integrated with community efforts in the future.

The third major finding describes a number of tactical issues related to technology and communications and accurately points out that planning and coordination among federal, state, and local public agencies are necessary in order to work together to resolve communications interoperability issues. To that end, the Wildland Fire Leadership Council has commissioned the development of a National Wildland Fire Enterprise Architecture team to improve interagency information technology and business practices. One of the focus areas for this effort will be geographic information systems used in wildland fire management by federal, state, tribal, and local agencies.

Following are several specific comments to the draft report that we hope you will find useful in producing the final report.

Now on p. 8.

See comment 1.

Page 10, Paragraph 1:  Research by Jack Cohen, et al. (USDA Forest Service) indicates home ignitions are not caused by radiant heat generated from passing crown fires or flame fronts.  Rather, showers of burning embers occur and are carried upwards via convection from fires often some distance away.  Collecting in a fashion similar to snow drifts, piles of embers accumulate on areas such as flammable roofs, corners where decks and wooden walls meet; on flammable adjacent objects like patio furniture cushions; under decks where leaves and other materials have collected; or they cascade and accumulate into vent openings in attics, igniting flammable materials within.  Structures often ignite several hours after the fire has passed, as these small piles of embers heat ignitable surfaces over time, finally causing an ignition.

Clearing a defensible space zone, and using less flammable construction materials helps minimize the number of ignitable surfaces, so when embers do accumulate there is less probability they will cause an ignition.  Surface or crown fires certainly can ignite a structure, if fuels are actually touching or closely adjacent to a structure.  With a modest clearance zone, radiant heat should be less of a factor.

Now on p. 12.

Page 14, Paragraph 2:  Homeowners are especially concerned about costs related to improving the fire survivability of their homes.  A fire-resistant roof, Firewise landscaping, and improved water access and delivery are some of the more effective and cost efficient measures for homeowners to take.  Double-pane windows are a prudent requirement, for new homes, not only for fire protection but for energy efficiency.  There is also a bit of controversy over whether protective gels and foams are effective or if it is prudent for citizens to take time to apply them when they could either be evacuating in a timely manner or engaging in other, possibly more effective, activities if they do choose to stay (i.e. removing vegetation, extinguishing small spot fires, etc.).  On the other hand, foams have been effectively used by professionally trained firefighters with specialized application equipment.

Now on p. 13.

Page 15:  It is true that many homeowners believe fire officials are responsible for protecting their homes.  Fully trained and highly equipped municipal fire departments responded to many of the historical WUI incidents cited in this report, with multiple alarms as well as with mutual aid providers from adjacent communities to assist.  In many of these incidents, and in the Oakland Hills fire specifically, these forces were overwhelmed by multiple structural ignitions, the resulting lack of water and pressure, poor access, and extreme fire behavior.  Less equipped rural and volunteer departments would certainly be overwhelmed by similar scenarios.  Firefighters will often be unable to access, much less protect, most individual homes, particularly in a high-density housing, multiple-ignition scenario, hence the need for wise preparation by individual homeowners.

Now on p. 14.

Page 16 Paragraph 2:  The report's assertion, that insurance companies have historically not placed emphasis on wildland fire in the past, is correct.  While there had been some early pilot efforts by individual companies in the intermountain West, in general, industry representatives in California have said that losses from the 2003 fires have prompted concern and the contemporary efforts to now more vigorously enforce various requirements, such as defensible space and fire resistant roofing.

Now on p. 43.

Page 46: The University of Nevada Cooperative Extension has done extensive research on factors that affect homeowner motivation to take preventive measures to protect their properties.

Now on p. 45.

Page 49: Community involvement is essential for the success of Firewise Communities programs. Even with ample funding, these programs will not be effective without community and individual participation. Governments, universities, nonprofits and other organizations can only provide the tools to initiate change. Personal responsibility is an essential theme in ongoing education efforts.

Now on p. 58.
See comment 2.

Page 62: The last sentence of footnote 46 should be changed to say, "These radios are routinely used for large fires and...." to properly state the common use of the radios.

Now on p. 66.
See comment 3.

Page 71: The National Interagency Coordination Center records a figure closer to 500 aircraft for the number of tactical and support aircraft utilized annually, instead of the 800 figure cited in footnote 59 of the report.

Now on p. 91.
See comment 4.

Page 96: The second to last sentence on the page should be changed to say, " VHF (AM and FM) is used by personnel and tactical aircraft on and over the fire line for tactical communications and UHF (AM) is used in the base camp for logistical, or other non tactical uses".

Thank you for the opportunity to review this draft report which provides a very useful overview of technology issues surrounding structure protection and communications during wildland fires. We look forward to the final report.

Sincerely,

P. Lynn Scarlett
**Acting** Assistant Secretary
Policy, Management and Budget

The following are GAO's comments on the letter from the Department of the Interior, dated April 12, 2005.

## GAO Comments

1. Crown fires can threaten structures if adequate defensible space is not present. In such cases, the flames from a crown fire can come into contact with a structure or the heat from the fire can damage a structure even without direct contact. Taking the protective measures discussed in our report—creating and maintaining defensible space and using fire-resistant roofs and vents—will reduce the risk of damage or destruction from wildland fire threats.

2. We revised the text to reflect that the radios in the cache are routinely used for large fires.

3. According to officials at the National Interagency Fire Center, the Forest Service and the Department of the Interior have a fleet of approximately 700 aircraft, including both large and small fixed-wing aircraft and helicopters. These include both government-owned and contracted aircraft. We have revised the text to reflect this information.

4. We revised the text to reflect that VHF (AM and FM) is used for tactical communications by federal firefighting personnel on the fire line and by tactical aircraft flying over the fire and UHF (AM) is used in the base camp for logistical, or other nontactical uses.

# GAO Contacts and Staff Acknowledgments

## GAO Contacts

Robin Nazzaro (202) 512-3841; nazzaror@gao.gov
Keith Rhodes (202) 512-6412; rhodesk@gao.gov
Steve Secrist (415) 904-2236; secrists@gao.gov
Naba Barkakati (202) 512-4499; barkakatin@gao.gov

## Staff Acknowledgments

In addition to the individuals named above, Dave Bixler, William Carrigg, Ellen W. Chu, Jonathan Dent, Janet Frisch, Robert Hadley, Barry T. Hill, Nicholas Larson, Kim Raheb, and Jena Sinkfield made key contributions to this report. Also contributing to this report were Michael Armes, Mark Braza, Joyce Evans, Timothy Guinane, Richard Hung, Chester Joy, Doug Manor, Cynthia Taylor, and Amy Webbink.

# Related GAO Products

**Previous Technology Assessments**

*Technology Assessment: Cybersecurity for Critical Infrastructure Protection.* GAO-04-321. Washington, D.C.: May 28, 2004.

*Technology Assessment: Using Biometrics for Border Security.* GAO-03-174. Washington, D.C.: November 15, 2002.

**Selected GAO Products Related to Wildland Fire Management**

*Wildland Fire Management: Important Progress Has Been Made, but Challenges Remain to Completing a Cohesive Strategy.* GAO-05-147. Washington, D.C.: January 14, 2005.

*Wildland Fires: Forest Service and BLM Need Better Information and a Systematic Approach for Assessing the Risks of Environmental Effects.* GAO-04-705. Washington, D.C.: June 24, 2004.

*Federal Land Management: Additional Guidance on Community Involvement Could Enhance Effectiveness of Stewardship Contracting.* GAO-04-652. Washington, D.C.: June 14, 2004.

*Wildfire Suppression: Funding Transfers Cause Project Cancellations and Delays, Strained Relationships, and Management Disruptions.* GAO-04-612. Washington, D.C.: June 2, 2004.

*Forest Service: Information on Appeals and Litigation Involving Fuel Reduction Activities.* GAO-04-52. Washington, D.C.: October 24, 2003.

*Geospatial Information: Technologies Hold Promise for Wildland Fire Management, but Challenges Remain.* GAO-03-1047. Washington, D.C.: September 23, 2003.

*Wildland Fire Management: Additional Actions Required to Better Identify and Prioritize Lands Needing Fuels Reduction.* GAO-03-805. Washington, D.C.: August 15, 2003.

*Wildland Fire Management: Reducing the Threat of Wildland Fires Requires Sustained and Coordinated Effort.* GAO-02-843T. Washington, D.C: June 13, 2002.

*Wildland Fire Management: Improved Planning Will Help Agencies Better Identify Fire-Fighting Preparedness Needs.* GAO-02-158. Washington, D.C.: March 29, 2002.

*Severe Wildland Fires: Leadership and Accountability Needed to Reduce Risks to Communities and Resources.* GAO-02-259. Washington, D.C.: January 31, 2002.

*The National Fire Plan: Federal Agencies Are Not Organized to Effectively and Efficiently Implement the Plan.* GAO-01-1022T. Washington, D.C.: July 31, 2001.

*Reducing Wildfire Threats: Funds Should be Targeted to the Highest Risk Areas.* GAO/T-RCED-00-296. Washington, D.C.: September 13, 2000.

*Western National Forests: A Cohesive Strategy Is Needed to Address Catastrophic Wildfire Threats.* GAO/RCED-99-65. Washington, D.C.: April 2, 1999.

## Selected GAO Products Related to Communications Interoperability and Spectrum Management

*Homeland Security: Federal Leadership and Intergovernmental Cooperation Required to Achieve First Responder Interoperable Communications.* GAO-04-740. Washington, D.C.: July 20, 2004.

*Spectrum Management: Better Knowledge Needed to Take Advantage of Technologies That May Improve Spectrum Efficiency.* GAO-04-666. Washington, D.C.: May 28, 2004.

*Project SAFECOM: Key Cross-Agency Emergency Communications Effort Requires Stronger Collaboration.* GAO-04-494. Washington, D.C.: April 16, 2004.

*Homeland Security: Challenges in Achieving Interoperable Communications for First Responders.* GAO-04-231T. Washington, D.C.: November 6, 2003.

*Telecommunications: Comprehensive Review of U.S. Spectrum Management with Broad Stakeholder Involvement Is Needed.* GAO-03-277. Washington, D.C.: January 31, 2003.